Mechanical Efficiency of Heat Engines

This book presents a newly developed general conceptual and basic quantitative analysis of the mechanical efficiency of heat engines. The book presents a theory of mechanical efficiency at a level of ideality and generality compatible with the treatment given to thermal efficiency in classical thermodynamics. This yields broad bearing results concerning the overall cyclic conversion of heat into usable mechanical energy. Most notably, the work reveals intrinsic limits on the overall performance of reciprocating heat engines. The ideal Stirling engine is shown to have the best overall performance potential of all possible engines operating under comparable conditions, which leads to mathematically explicit universal upper bounds on mechanical efficiency and cyclic work output. The theory describes the general effects of parameters such as compression ratio and external or buffer pressure on engine output. It also provides rational explanations of certain operational characteristics such as how engines generally behave when supercharged or pressurized. The results also identify optimum geometric configurations for engines operating in various regimes from isothermal to adiabatic. The basic mechanical efficiency results are extended to cover multi-workspace engines and heat pumps. Limited heat transfer and finite-time effects have also been incorporated into the work.

The main research interests of Prof. James Senft lie in the mathematical analysis of mechanisms and heat engines, with an emphasis on the Stirling engine. He has published more than 40 papers and several books in these areas. He holds the position of Professor Emeritus of Mathematics at the University of Wisconsin–River Falls. He has been a visiting research professor at the University of Rome, the University of Washington Joint Center for Graduate Study, the University of Zagreb, and the University of Calgary. Professor Senft has been a visiting fellow at the Australian National University Institute of Advanced Study and a visiting scientist at Argonne National Laboratory. He has received research grants from the Charles A. Lindbergh, Fulbright, and the National Science Foundations and has served as a consultant to the U.S. Department of Energy and NASA.

MECHANICAL EFFICIENCY OF HEAT ENGINES

James R. Senft

University of Wisconsin–River Falls

CAMBRIDGE UNIVERSITY PRESS

CAMBRIDGE UNIVERSITY PRESS
Cambridge, New York, Melbourne, Madrid, Cape Town, Singapore,
São Paulo, Delhi, Dubai, Tokyo, Mexico City

Cambridge University Press
The Edinburgh Building, Cambridge CB2 8RU, UK

Published in the United States of America by Cambridge University Press, New York

www.cambridge.org
Information on this title: www.cambridge.org/9780521169288

First published 2007
First paperback edition 2010

A catalogue record for this publication is available from the British Library

Library of Congress Cataloging-in-Publication Data

Senft, J. R. (James R.), 1942–
Mechanical efficiency of heat engines / James R. Senft.
    p.   cm.
Includes bibliographical references and index.
ISBN-13: 978-0-521-86880-8 (hardback)
ISBN-10: 0-521-86880-7 (hardback)
1. Heat-engines.   2. Mechanical efficiency.   3. Thermodynamics. I. Title.
TJ255.S36   2006
621.402'5 – dc22   2006022394

ISBN 978-0-521-86880-8 Hardback
ISBN 978-0-521-16928-8 Paperback

This book is dedicated
to my son
VICTOR
the engineer.

CONTENTS

PREFACE

This book presents a general conceptual and basic quantitative analysis of the mechanical efficiency of heat engines. Typically, treatment of the mechanical efficiency of heat engines has been performed on a case-by-case basis. In ordinary practice, kinematic analysis and computer simulation of specific engine mechanisms coupled with calculated or measured pressure–volume cycles usually can indeed be effectively used for evaluating and locally optimizing engine designs. However, going beyond the specific and local requires broader insights that only a general theory can provide.

No general approach to mechanical efficiency of heat engines had been available until recently. This is in sharp contrast to the situation regarding the thermal efficiency of heat engines. Classical thermodynamics treats the subject of thermal efficiency in great generality. Its results, although obtained in a highly idealized setting, are of profound importance to engine theorists, designers, and practitioners. This book presents a theory of mechanical efficiency at a similar level of ideality and generality.

The first results in this area were published in 1985 and further developed in a series of papers up to the writing of this book. The work modeled the interaction between the mechanical section of an engine and its thermal section at a level compatible with that of classical thermodynamics. This yielded results of broad bearing concerning the overall cyclic conversion of heat into usable mechanical energy.

Most notably, the work uncovered intrinsic limits on the overall performance of reciprocating heat engines. The ideal Stirling engine was shown to have the potentially best overall performance of all possible engines operating under comparable conditions. The work provided mathematically explicit upper bounds on the mechanical efficiency and cyclic work output of engines having like characteristics. The theory described the effects of parameters such as compression ratio and external or buffer pressure on engine output. It also provided rational explanations of certain operational characteristics such as how engines generally behave when supercharged or pressurized. The results also identified optimum geometric configurations for engines operating in various regimes, from isothermal to adiabatic. Limited heat transfer effects have also been incorporated into the work, and results have been extended to multiworkspace engines and heat pumps as well. Most of this has been collected, organized, and presented in the pages that follow.

The research reported in these pages began during a stay as a visiting scientist at Argonne National Laboratory, was subsequently supported by three grants from the National Science Foundation, and was aided by visiting professorships at the University of Rome and the Australian National University. My own University of Wisconsin in River Falls sustained the work throughout and provided a sabbatical which made most of the compilation of this book possible. I am grateful for all of the opportunities given to me.

In addition to these institutional benefactors, a certain small group of people must be specially thanked for the completion of this book. These poor souls have been afflicted with my presence in their lives, and they each in their own way provided encouragement and motivation to me for continuing with the writing of this book especially at times when I did not want to.

First and foremost of this group is my wife, Gloria, who, having patiently suffered though all of my ups and downs for some 39 years, was particularly tried during the writing of this book. She never wavered in her belief that my duty was to finish the book I had been given to do.

Many colleagues at the university were also steadfastly supportive, especially David Yurchak and Keith Chavey, who served as chairs of the department through this ordeal, and Kevin McLaughlin of the Chemistry Department who helped by his constant cheerful interest.

I am grateful to Fr. John Beckfelt, pastor of St. Mary's Parish in Big River, who regularly supplied encouragement and wise advice. I thank all the Saints whom I asked for prayers on my behalf . . . and above all, I thank our good God who hears all prayer.

JRS

As God is above all created things, honors, and possessions,
so should our internal esteem of his Divine Majesty
surpass our esteem or idea of anything whatever.

Saint Aloysius Gonzaga

1

ENERGY TRANSFERS IN CYCLIC HEAT ENGINES

Heat engines are made to provide mechanical energy from thermal energy. Efficiency is a convenient measure of how well this is done. The overall efficiency of an engine is usually thought of as the product of two more basic efficiencies: the thermal efficiency of the engine cycle and the cyclic mechanical efficiency of the complete device. The first is well treated in classical thermodynamics. The second, mechanical efficiency, is the subject of this work.

Analysis of the mechanical efficiency of heat engines can be only as general as the conceptual basis on which it is built. The model used here has a level of generality matching that used in classical thermodynamics to analyze the thermal efficiency of heat engines.

HEAT ENGINE DIAGRAMS

Figure 1.1 is a representation of a cyclic heat engine typically found in thermodynamics textbooks. G represents the body of the working substance, and T_H and T_C are the temperatures of the heat source and sink, respectively. The net or *indicated cyclic work* done by the engine working substance is the difference $Q_i - Q_o = W$ between the heat absorbed from the high temperature source and the heat rejected to the lower temperature reservoir during a complete cycle.

Although the diagram is adequate for discussing the *thermal efficiency* $\eta_t = W/Q_i$ of the cycle, it does not allow the analysis of all of

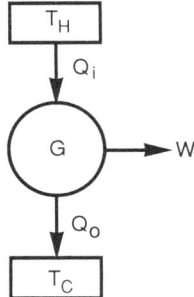

Figure 1.1 Cyclic heat engine diagram depicting heat transfers Q_i and Q_o to and from the working substance and indicated work output W.

the mechanical energy transfers that determine the mechanical efficiency of a complete engine. In fact, work must be done on the engine fluid to carry out half the cycle.

This is quite clear in looking at any pressure–volume $(p-V)$ diagram of an engine cycle. An example is given in Figure 1.2 with characteristics that are typical of the cycles encountered in elementary thermodynamics and normal practice. What is termed as a *regular cycle* is described by a pair of functions p_c and p_e defined and continuous on a closed bounded interval $I = [V_m, V_M]$ representing the volume

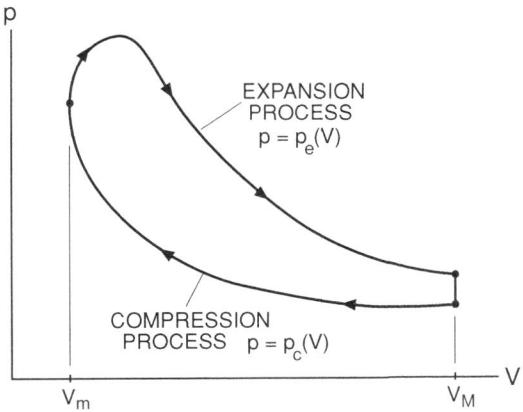

Figure 1.2 A regular cycle in the $p-V$ plane.

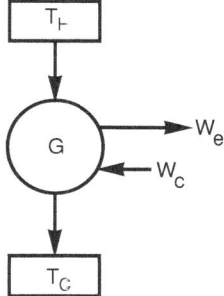

Figure 1.3 Heat engine diagram showing work transfers W_c and W_e with the working substance.

variation of the cycle. The functions represent the compression and expansion pressures of the cycle, with vertical segments supplied at volume extremes if necessary. For an engine, the cycle is oriented as shown in the figure, whereas compressors or heat pumps have the opposite orientation; the discussion will be limited to engines until a later chapter.

The area enclosed by the cycle in the $p–V$ plane is the indicated work W of the engine. This net cyclic work is the difference between two distinct work processes. It is the difference between the work done by the engine fluid during expansion and the work done on it during compression. The *absolute expansion work* of the cycle is the area directly under the upper curve $p = p_e(V)$:

$$W_e = \int_{V_m}^{V_M} p_e(V)\,dV. \tag{1.1}$$

The *absolute compression work* of the cycle is the area directly below the cycle, which represents work that must be done on the engine substance to carry out the compression process described by the lower curve $p = p_c(V)$:

$$W_c = \int_{V_m}^{V_M} p_c(V)\,dV. \tag{1.2}$$

Both work quantities as defined here are positive, and $W = W_e - W_c$.

Figure 1.3 shows the individual expansion and compression work transfers. In a reciprocating or cyclic working engine, the expansion and compression processes do not take place simultaneously but rather sequentially. Thus to realize a self-acting reciprocating engine, means must be provided to divert and store some of the absolute expansion work and redirect it to the engine working fluid when it needs to carry out its absolute compression work.

THE BASIC CYCLIC HEAT ENGINE

Most practical engines have the features depicted conceptually in Figure 1.4. This is the type of engine to be dealt with here and will be referred to as a *reciprocating* or *cyclic kinematic* engine. The working substance, typically a gas, is contained in a capsule called the *workspace*, which is equipped with a means for varying the volume, usually a *piston*. Only a single body of working gas will be considered for the present. This does not represent a significant loss of applicability because most multi-cylinder engines can be considered as parallel connections of single-workspace engines. A later chapter will treat more complex arrangements of multi-piston engines.

The workspace is also equipped with means to interact thermally with heat reservoirs (not shown in Figure 1.4). The prime characteristic of the reciprocating engine is the *mechanism* linking the piston to the output *shaft*. This link is a kinematic one. The motion of the piston and

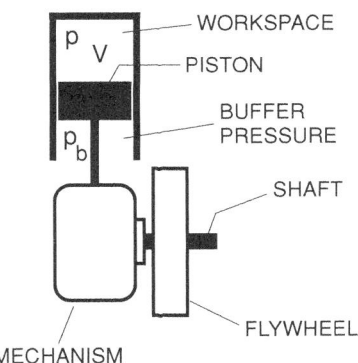

Figure 1.4 The elements of a reciprocating heat engine.

all other moving parts of the engine is completely constrained by the mechanism. The mechanism transmits force or torque as well as motion, so it is actually a *machine* in proper parlance, but the term *mechansim* will be used here to help avoid confusion with the engine as a whole.

The workspace usually contains other devices not shown in the figure such as valves or displacers or whatever may be necessary to carry the working fluid through the desired thermodynamic cycle. These devices, as well as auxiliary pumps, fans, etc., are kinematically linked to and are driven by the mechanism and are conceptually considered as part of the mechanism in the analysis here.

In the turbine type of heat engine, the expansion and compression processes take place simultaneously in different locations in the engine, and the processes are continuous. In cyclic kinematic engines, the processes are discrete and sequential. Because of this, a kinematic engine must be equipped with at least one work reservoir. For single-workspace engines with a rotating shaft output, this reservoir invariably takes the form of a *flywheel*, as Figure 1.4 depicts. Other devices can be used such as pendulums or springs at appropriate places. In multi-workspace engines, each workspace can use some of the others for this purpose as well.

Under steady state operation, the flywheel does not experience a net gain in energy over a cycle. During each cycle, it absorbs, stores, and returns energy to the engine that is necessary to sustain the cycle; the remainder is directed through the output shaft for use outside the engine.

BUFFER PRESSURE

The single-workspace engine needs nothing more in principle than the features described, but in practice it has a near constant external *buffer pressure* acting on the non-workspace side of the piston. The source of this pressure is usually due to the surrounding atmosphere, but sometimes a special enclosure or *buffer space* is constructed to permit the use of elevated pressure. As will be shown, buffer pressure

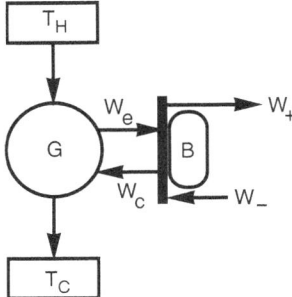

Figure 1.5 An engine diagram showing the effect of buffer pressure on piston work transfers.

has a significant influence on the mechanical efficiency of a kinematic engine.

All work transferred through the engine mechanism is subject to some loss due to friction. This applies to transfer in both directions: from flywheel to piston as well as from piston to flywheel and output shaft. The works transferred are not generally W_e and W_c because of the action of the buffer pressure. The buffer gas, like the flywheel, absorbs, stores, and returns energy to the working gas during the cycle. But it acts directly on the piston and thus diverts and recycles some work duty away from the mechanism. Figure 1.5 represents the buffer space with the element labeled B. The work quantities that must be mechanically transmitted to and from the engine piston are reduced from W_e and W_c by the influence of the buffer gas pressure to what will be denoted by W_+ and W_-. This reduces the friction losses in the mechanism section of the engine in a way to be precisely described shortly. W_+ and W_- will be referred to as the *efficacious* and *forced piston work*, respectively. Conceptually, W_+ is the non-negative work done on the mechanism by the piston in each cycle; W_- is the non-negative work done on the piston by the mechanism in each cycle. Note that both of these quantities are non-negative by definition. Also take note of the fact that the arrows in Figure 1.5 refer to positive work transfer only; they do not necessarily correspond to the direction of piston motion or to the direction of force or torque.

Since buffer presure acts directly on the piston, it is not subject to the frictional losses that energy storage in the flywheel entails. Any piston seal friction present is most appropriately included with the friction of the mechanism section and not associated with the buffer gas. The buffer gas, however, may suffer loss in another way when the buffer space volume is finite. This loss is often termed *hysteresis* or *transient heat transfer loss* (West, 1986). It occurs when the pressure of the buffer space fluctuates with the volume changes induced by the piston motion. In the interior of the buffer space, the gas experiences a corresponding temperature fluctuation. The walls of the buffer space are nearly isothermal. This produces a net flow of energy from the gas to the container walls over each cycle. To counter this in practice, one makes the buffer space volume as large as practical. This minimizes the pressure excursion and therefore minimizes the hysteresis loss. Usually the pressure variation can be kept relatively small, and therefore it is reasonable in most analyses to assume the ideal of constant buffer pressure. *Constant buffer pressure will be our usual assumption* in all of the following work except where noted, as in Chapter 8. If the pressure is constant, the buffer gas acts as a lossless energy reservoir in direct communication with the piston. In this ideal case the difference between the efficacious and forced work is exactly the indicated work of the cycle:

$$W_+ - W_- = W = W_e - W_c .$$

Other devices can be arranged to act on the piston directly, such as springs or weights, but assuming a constant buffer pressure is adequate to cover most applications.

SHAFT WORK

The complete cyclic kinematic heat engine as just described can be conceptually represented by the diagram of Figure 1.6. Device M represents

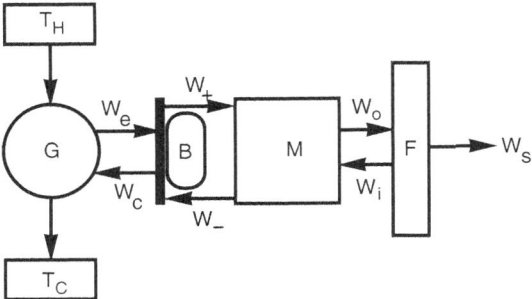

Figure 1.6 Diagram showing the work transfers between the basic components of a cyclic kinematic heat engine.

the mechanism and F the flywheel. The quantity W_s is the *cyclic shaft work*. In practical terms it is the useful work coming out of the engine in each cycle. In the diagram, it is the difference between the cyclic work W_o received by the flywheel/output shaft from the mechanism and the cyclic work W_i taken from the flywheel and directed into the mechanism to sustain operation: $W_s = W_o - W_i$.

W_o is the efficacious piston work W_+ reduced by friction losses in passing through the mechanism. W_i is the work that the flywheel must feed into the mechanism in order to deliver, after having been reduced by friction losses in passing through the mechanism, the work W_- that the piston is forced to do. The effectiveness of the mechanism in transmitting all this work determines the difference between the indicated work of the engine cycle and the shaft work that is ultimately delivered:

$$W_o \leq W_+ \quad \text{and} \quad W_i \geq W_- \quad \text{make} \quad W_s = W_o - W_i \leq W_+ - W_- = W .$$

A subtle point, however, is that the shaft work also depends on how much energy the mechanism must transmit. How much is determined by the positive and negative piston works. The magnitude of these quantities are dependent not only upon the compression and expansion works of the cycle, W_c and W_e, but also depend significantly upon cycle shape and buffer pressure.

BUFFER PRESSURE AND ENERGY TRANSFERS

Figure 1.7 illustrates how buffer pressure level determines where and when efficacious and forced piston work occurs. An elliptical p–V cycle is shown with an increasing sequence of constant buffer pressure levels. A plus sign along a segment of the cycle indicates efficacious piston work, i.e., where positive work is done on the mechanism by the piston. This occurs during those portions of the cycle where either the workspace pressure is above the buffer pressure and the piston is effecting a volume expansion, or where the workspace pressure is below the buffer pressure and a compression is taking place. A minus sign signifies the opposite situation, namely, a process in which work must be taken from the flywheel and transmitted to the piston by the mechanism. Note that the signs do not necessarily relate to the direction of piston motion but rather denote the direction of positive work transfer between the piston and the mechanism.

These energy exchanges and the corresponding mechanical losses will be quantified in the following chapters, but some important and useful points should be intuitively clear at this juncture. First, an engine buffered as in Figure 1.7(a) will require more flywheel effect than that in (b). Case (c) will require even less, and (d) will need minimal flywheel assistance. Accordingly, larger mechanical losses appear more likely in Case (a) than in (b) and so on to (d). This will in fact be shown to be the case under very general assumptions in the next chapter. Higher buffer pressures as in (e), (f), and (g) have effects similar to the lower pressures.

It should also be noted that no choice of constant buffer pressure will entirely eliminate the need for a flywheel for this particular cycle. At any buffer pressure level there will be some segment labeled with a minus sign, and therefore some stored flywheel energy will be needed. This is typical for most cycles but is not always the case. For example, in each cycle in Figure 1.8 there is at least one buffer pressure level that results in only work transfer from the piston to the output shaft and no forced piston work occurs.

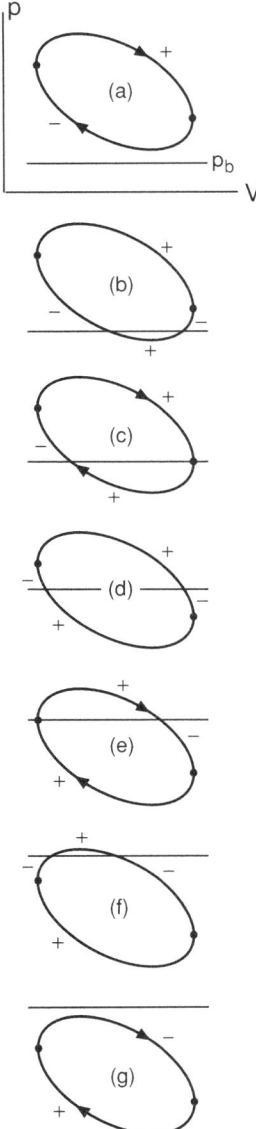

Figure 1.7 An elliptical engine cycle shown with various buffer pressure levels. The signs indicate the direction of energy transfer between the piston and the mechanism.

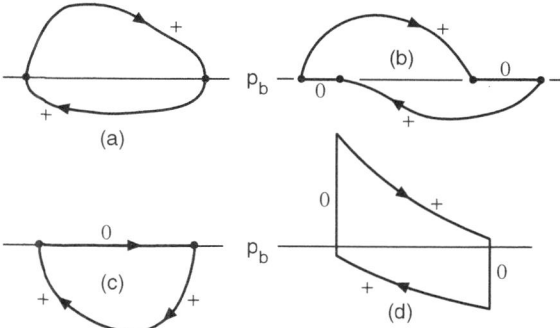

Figure 1.8 Examples of regular cycles which have no forced work for certain buffer pressure levels.

The cycle in Figure 1.8(a) has the buffer pressure line passing exactly through the volume extreme points, and so all segments are labeled with a plus sign. In Figure 1.8(b), the only non-plus segments occur where the cycle pressure equals the buffer pressure, and these are labeled with a zero. Since there is no pressure difference across the piston in these locations, ideally no force need be applied to the piston, so no work transfer takes place. Figure 1.8(c) is the imagined cycle of a vacuum engine in which hot gas is drawn into a cylinder at atmospheric pressure and then captured by a closing valve and cooled; the pressure drops and a compression follows in which energy is transfered to the engine shaft. Figure 1.8(d) is an ideal Stirling cycle in which constant volume processes occur at the volume extremes. The general convention is adopted that no work transfers take place on vertical segments since no piston motion occurs there. In practice, other engine parts, such as displacers in Stirling engines, may be in motion during dwell periods of the piston, and with that there is some associated friction loss. However, such losses are most appropriately included with mechanism losses falling under the same general category as valve train, seal friction, and auxiliary devices. When vertical segments do occur in a cycle, there may be a range of buffer pressure values for which $W_- = 0$, as is the case in Figure 1.8(d).

2

MECHANISM EFFECTIVENESS
AND MECHANICAL EFFICIENCY

In a cyclic heat engine, the mechanism plays a key and complicated role. Its main objective is to transport energy from the working substance to the output shaft. But it also functions to constrain and effect the movement of the piston in order that it carry out a certain thermodynamic cycle. This requires that the mechanism work in a bidirectional fashion. It must transport work from the piston to the flywheel and output shaft during some parts of the cycle, and from the flywheel to the piston in other parts. In practice, it is sometimes even more complex. For example, just after dead center in some engines, both the piston and the flywheel supply work to the mechanism, which is consumed by friction.

For analytic treatment, a comprehensive model of machines that reflects in detail all of the modes in which a mechanism is called upon to function in an engine is the natural first thought. However, such a model quickly becomes exceedingly complex, as the development in Appendix A shows. Rather, the main text of this monograph employs only very basic principles and examines best possible cases. As will be seen as the chapters unfold, a surprising number of interesting and practical insights about ultimate engine performance can be easily deduced through this simple approach.

MECHANISM EFFECTIVENESS

Consider first a point in the operation of an engine in which the working gas is expanding against the outwardly moving piston. If the engine pressure is above the buffer pressure, the net gas force on the piston is applied to the mechanism and a corresponding torque is present at the shaft. The ratio of this torque to the ideal torque, which would be present if there were no friction in the mechanism, will be called the *effectiveness* of the mechanism. This is simply the instantaneous version of the usual elementary textbook definition of the efficiency of a machine.[†] Using the term *effectiveness* throughout our analyses, however, will help emphasize and isolate the role that the performance of the mechanical section of an engine plays in the mechanical efficiency of the whole engine over a complete cycle.

Mechanism effectiveness in principle depends in a complex way upon a number of variables. It obviously depends upon the instantaneous position of the parts of the mechanism, which determines the loading on the various joints and hence the acting Coulomb friction forces. Inertial effects due to the velocity and acceleration of parts with appreciable mass also affect the joint loads and friction. Mechanism effectiveness may also depend significantly upon the magnitude of the force applied to the piston as when a friction type other than Coulomb is present in some joints.

If an engine piston is moving inward on a compression process in which the workspace pressure is below the buffer pressure, the situation is similar to the expansion process described above. The mechanism can transfer work from the piston to the output shaft, and instantaneous effectiveness would be determined in the same way. However, if a compression is occurring when the buffer pressure is below the workspace

---

[†] A technical definition of effectiveness in a general setting is given in Appendix A.

pressure, then the situation becomes inverted. The flywheel must supply work to the mechanism to be transferred to the piston. The shaft becomes a torque input device and the piston is where the output force appears. The effectiveness of the mechanism in this type of process is the ratio of the force that actually appears at the piston to the ideal force that would be there if no friction occurred in the mechanism; it is, of course, the exact same measuring scheme with input and output sides reversed. The case in which an expansion is forced to take place when the workspace pressure is below the buffer pressure is treated the same way.

This explains exactly how the term *mechanism effectiveness* will be understood in the analyses that follow. Further, it will be assumed throughout most of this work, unless explicitly stated otherwise, that the mechanism serves only as a work transducer and not as a work reservoir. That is, during parts of a cycle the mechanism itself does not accept and store any appreciable amount of energy and release it in other parts of the cycle. This models most practical engines sufficiently well in two ways. First, any energy stored internally as the kinetic energy of moving links and other parts, or in the position of internal springs or weights, is usually small compared to that stored in the flywheel or buffer gas, which are treated here as separate from the mechanism. Second, if mechanism parts such as, for example, a massively proportioned or very high speed crankshaft do store significant amounts of energy, they can usually be conceptually and quantitatively associated with the flywheel. Separating the function of the mechanism from the flywheel, or an equivalent external energy storage device, not only reflects typical practical situations, but also allows insights into the performance of an engine as a function of simple measures of the performance of its main components.

Clearly, mechanism effectiveness as described above is a nonnegative quantity and cannot exceed unity. There may in fact be portions of an engine cycle where the effectiveness is actually zero.

This is the case in the situation described earlier where both piston and flywheel put work into the mechanism in certain parts of its cycle.

MECHANICAL EFFICIENCY

The *mechanical efficiency* of an engine measures how much of the work produced by the thermodynamic cycle of the workspace can actually be taken from the shaft for use outside the engine:

$$\eta_m = \frac{W_s}{W} \tag{2.1}$$

where W_s is cyclic shaft work and W is indicated cyclic work. Note well that mechanical efficiency is a cyclic quantity, determined over a complete cycle of operation of the engine.

In the previous chapter it was shown that the difference between W and W_s centers about the efficacious and forced piston work of the engine cycle and buffer pressure. The efficacious work W_+ is work input by the piston to the mechanism for transmittal to the flywheel shaft. In transit, some is lost to friction and the reduced amount W_o is actually delivered. As illustrated at the beginning of this chapter, determining this reduced amount requires complex analysis. Suppose however, that, as is often the case in practice, constant bounds on the instantaneous mechanism effectiveness ε are known or can be surmised. Suppose it is known that the effectiveness is at least $L > 0$ and at most $H \leq 1$ throughout the cycle. Then from the definition of mechanism effectiveness given above, it follows that

$$L W_+ \leq W_o \leq H W_+ .$$

The forced work W_- is work that the mechanism must deliver to the piston to make it move in opposition to the pressure difference across it. Because of losses in transmission through the mechanism, more work, namely W_i, must be taken from the flywheel to supply W_-.

Again using the same effectiveness bounds, the definiton of mechanism effectiveness implies

$$L W_i \leq W_- \leq H W_i .$$

Combining these two inequalities with $W_s = W_o - W_i$ yields

$$L W_+ - \frac{W_-}{L} \leq W_s \leq H W_+ - \frac{W_-}{H} . \qquad (2.2)$$

Because $W = W_+ - W_-$, Inequality (2.2) can be put into the form

$$L W - \left(\frac{1}{L} - L\right) W_- \leq W_s \leq H W - \left(\frac{1}{H} - H\right) W_- \qquad (2.3)$$

or its equivalent in view of definition (2.1) by dividing through by W:

$$L - \left(\frac{1}{L} - L\right) \frac{W_-}{W} \leq \eta_m \leq H - \left(\frac{1}{H} - H\right) \frac{W_-}{W} . \qquad (2.4)$$

This result (Senft, 1985) indicates the important role that forced work plays in determining the useful output of an engine. Its presence extracts a double reduction in the output of the engine, one reduction for each passage through the mechanism. This is in addition to the expected reduction of the indicated cycle work by the mechanism effectiveness.

The presence of any forced work always reduces mechanical efficiency to a value below that of the effectiveness of the mechanism. For example, suppose it is known that the effectiveness of the mechanism of a certain engine is between 0.8 and 0.9 over its cycle. Suppose that for a chosen buffer pressure the ratio of forced work to indicated work is 1/4. Then (2.4) shows that the mechanical efficiency of that engine over a complete cycle cannot actually exceed 0.85 and may be as low as 0.69:

$$0.69 = 0.8 - \left(\frac{1}{0.8} - 0.8\right) \frac{1}{4} \leq \eta_m \leq 0.9 - \left(\frac{1}{0.9} - 0.9\right) \frac{1}{4} = 0.85 .$$

Such applications can be useful for quick estimates in practical situations. It should be pointed out, however, that the lower bound often is not applicable because the effectiveness of many engine mechanisms drops completely to zero near dead center or piston reversal points. Hence, in most of what follows. use of the upper bound will dominate. As will be seen in the following chapter, many insights on general engine performance limits will follow from (2.3) and (2.4).

FORCED WORK

Given that forced work is the connection between mechanism effectiveness and mechanical efficiency, it is well to understand clearly what factors influence forced work. Figure 2.1 shows the engine cycle of Figure 1.2 with a particular choice of buffer pressure. The forced work is shown as the shaded area. For a regular cycle, forced work is analytically obtained by integrating to find the area that is below the buffer pressure line but above the expansion process pressure, and likewise the area that is above the buffer pressure line but below the compression process line.

As can be readily appreciated from this example, forced work depends upon the shape of the cycle and upon the buffer pressure level.

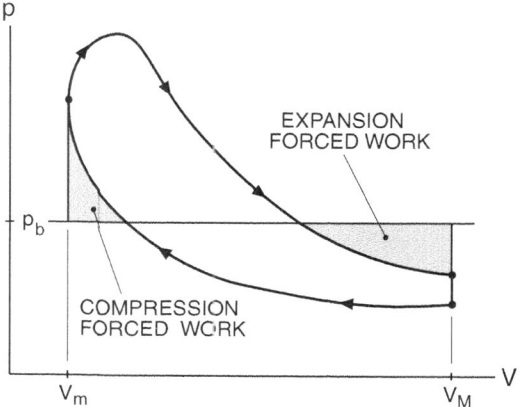

Figure 2.1 The shaded areas represent the forced work for this regular engine cycle with the constant buffer pressure level p_b.

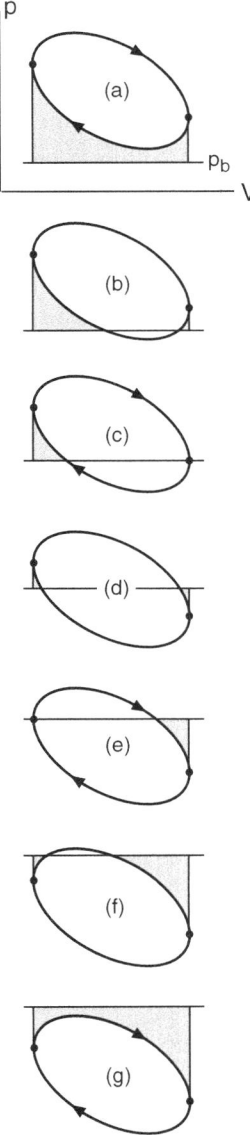

Figure 2.2 The influence of buffer pressure level on the forced work of an elliptical engine cycle.

These influences can be further understood by looking at the forced work for the elliptical cycle of Figure 2.2 with various buffer pressures. Note that forced work occurs only during the compression process for lower buffer pressures in (a)–(c). After this, some forced work appears on expansion as well as compression, but after buffer pressure is raised still higher, forced work occurs only during expansion.

3

GENERAL EFFICIENCY LIMITS

THE FUNDAMENTAL EFFICIENCY THEOREM

In this chapter, previous results are applied to deduce general upper bounds on the performance of large classes of engines characterized by a few simple parameters. The following theorem, which is a consequence of Inequality (2.4), is the main tool used (Senft, 1985, 1987a).

FUNDAMENTAL EFFICIENCY THEOREM

If the effectiveness ε of an engine mechanism is bounded above by the constant $0 < E \leq 1$, that is, if $\varepsilon \leq E$ throughout the cycle, then the mechanical efficiency η_m of the engine satisfies

$$\eta_m \leq E - \left(\frac{1}{E} - E\right)\frac{W_-}{W} \qquad (3.1)$$

where W is the indicated work of the cycle and W_- is the forced work. Equality holds in (3.1) if the mechanism effectiveness equals the constant E over the entire cycle and buffer pressure is constant.

The straightforward use of this theorem is to obtain an upper estimate on the mechanical efficiency of a proposed engine. For example, the engine cycle of Figure 3.1 is that of a beta type Stirling engine analytically obtained by use of the well-known Schmidt model (Senft, 2000). For the buffer pressure shown, the forced-to-indicated work ratio is about $0.30 = W_-/W$. If it were estimated or analytically determined that the engine mechanism had an effectiveness of 90% or less, then

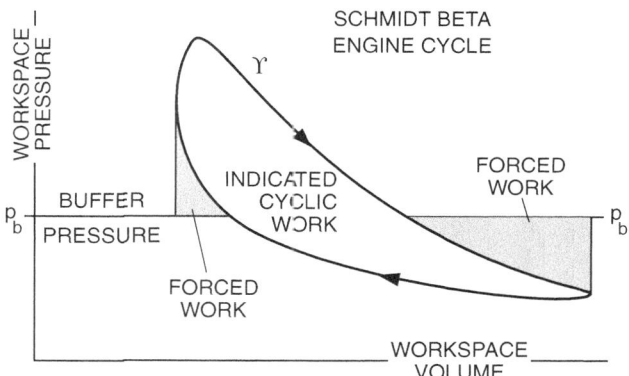

Figure 3.1 Schmidt pressure–volume diagram of a single-cylinder Stirling engine.

the theorem would show that the mechanical efficiency of the engine could not exceed 84%:

$$\eta_m \leq 0.90 - \left(\frac{1}{0.90} - 0.90\right)(0.30) = 0.837.$$

This is a substantial reduction from the 90% performance level of the mechanism. This is an important revelation presented by the theorem in general, namely, that *the mechanical efficiency of an engine over a cycle is always less than the effectiveness level of its mechanism when forced work is present.* The theorem also clearly shows that the performance potential in this example could be improved by lowering the buffer pressure to decrease the forced-to-indicated work ratio. Furthermore, if the effectiveness were everywhere equal to the constant E, then the mechanical efficiency of the engine would equal the upper bound given in the inequality. These observations will now be extended to make broad engine comparisons.

THE STIRLING COMPARISON THEOREM

Consider an arbitrary engine, called A, with a given cycle and buffer pressure. Denote its forced and indicated works by $W_-\langle A\rangle$ and $W\langle A\rangle$ respectively. Suppose for a given operating speed, or range of speeds,

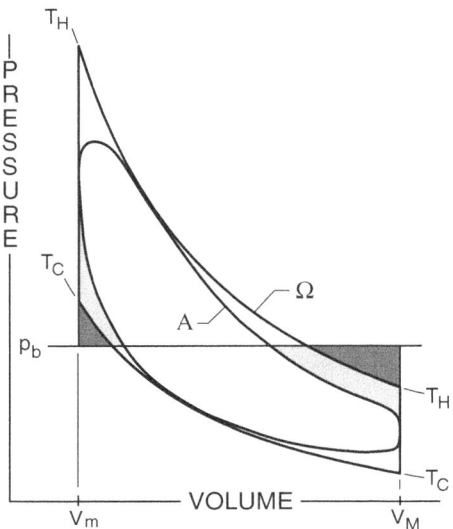

Figure 3.2 An engine cycle A inscribed in an ideal Stirling cycle Ω. The dark shaded area is the forced work of the Stirling cycle. The forced work of the inscribed cycle is the lighter shaded area plus the dark.

its mechanism effectiveness ε is bounded by the constant E, that is, $\varepsilon \leq$ E. Then the Fundamental Efficiency Theorem (3.1) applies, giving

$$\eta_m\langle A \rangle \leq E - \left(\frac{1}{E} - E \right) \frac{W_-\langle A \rangle}{W\langle A \rangle} \ .$$

Now the right-hand side of this inequality is the mechanical efficiency of an engine with the same cycle and buffer pressure as engine A, but with a mechanism – albeit hypothetical – having constant effectiveness E throughout its range of operation.

The given engine A and its hypothetical better have the same cycle, so they have the same volume extremes and the same temperature extremes. Consider now an ideal Stirling engine Ω also with the same volume and temperature extremes. If its cycle is taken to have the same mass of the same ideal gas as A, the cycles are related as in Figure 3.2, with cycle A inscribed within Ω. With the same buffer pressure outside both cycles, clearly $W_-\langle A \rangle \geq W_-\langle \Omega \rangle$. Since $W\langle A \rangle \leq W\langle \Omega \rangle$, it follows that

$$\frac{W_-\langle A \rangle}{W\langle A \rangle} \geq \frac{W_-\langle \Omega \rangle}{W\langle \Omega \rangle} \ .$$

Since $E \leq 1$, it further follows that

$$E - \left(\frac{1}{E} - E\right)\frac{W_-\langle A\rangle}{W\langle A\rangle} \leq E - \left(\frac{1}{E} - E\right)\frac{W_-\langle\Omega\rangle}{W\langle\Omega\rangle} \,.$$

Now the right-hand side of this, according to the Fundamental Theorem again, is the mechanical efficiency of the ideal Stirling engine Ω if equipped with a mechanism of constant effectiveness E. This shows that $\eta_m\langle A\rangle \leq \eta_m\langle\Omega\rangle$, and because the engine A was arbitrary, we have the following theorem (Senft, 1985, 1987a).

STIRLING COMPARISON THEOREM

Of all cyclic heat engines working between the same temperature extremes, and having the same volume extremes, the same mass of ideal gas, the same constant buffer pressure, and mechanisms with effectiveness E or less, the ideal Stirling with a mechanism of constant effectiveness E has the maximum mechanical efficiency.

The comparison made in this theorem is a fair and natural one. The engines being compared operate between the same source and sink temperatures. They have the same mechanism effectiveness potential E and work against the same buffer pressure. The engines are of comparable size inasmuch as they have the same maximum and minimum workspace volumes. Finally, they use the same mass of the same ideal gas. Comparing engine performance relative to their size, temperatures, working fluid, and mechanism merit is conceptually natural and fair. The theorem above can be loosely restated as *all engines in a fair comparison class have a mechanical efficiency not exceeding that of an ideal Stirling with a mechanism of constant effectiveness equal to the bound for the class*. Note that the ideal Stirling also has the maximum cyclic shaft work output of all engines in the class.

The theorem is not a simple tautology. The theorem states that in a comparison of engines within given natural limits of size, temperature extremes, gas fill, and mechanism merit, the ideal Stirling yields maximal performance. The proof of the theorem obviously makes clear that its validity does not rest solely upon abstract logical considerations. The proof relies upon the Fundamental Efficiency Theorem in an essential way. It is because of this theorem that the Stirling Comparison Theorem is true.

In cases where an engine A has a buffer pressure given for which there is no forced work, as in the examples in Figure 1.8, then the Fundamental Theorem trivially reiterates the obvious, namely, $\eta_m\langle A \rangle \leq E$. Cycle and buffer combinations having no forced work will often be referred to as *efficacious*. The Stirling Comparison Theorem also loses its force in an efficacious case. In this case, the Stirling covering cycle Ω is as pictured in Figure 1.8(d) with the buffer pressure line passing through both isometric parts of the cycle. With constant mechanism effectiveness E, $\eta_m\langle \Omega \rangle = E$, which simply gives $\eta_m\langle A \rangle \leq E$ again.

Although efficacious engines are the most efficient mechanically, they are not always the most practical. Non-efficacious engines potentially have a higher power density. Also, some engines, such as high compression two-stroke or four-stroke internal combustion engines cannot be rendered efficacious by any choice of external pressure, a topic which will be examined later.

CONSTANT MECHANISM EFFECTIVENESS

As the results of the previous section suggest, the case of engines with mechanisms of constant effectiveness is important from a theoretical point of view. There are no reciprocating engines in practical use having mechanisms of constant effectiveness. Indeed, the ubiquitous crank and connecting rod drive has a variable effectiveness which drops to zero near dead centers. It is, however, so remarkably constant over most of its motion between reversals (Senft, 1993b) that constant effectiveness is an appropriate approximation for many purposes.

One can conceive of constant-effectiveness engine mechanisms such as the reversing partial gear and double rack drives so favored by Ramelli (1588), but they seem to be suitable only for relatively slow-moving, low-power-density machines. However, the assumption of constant mechanism effectiveness is appropriate and fruitful for making first order approximations, and even more for developing theoretical understanding of best possible cases. The Fundamental Efficiency Theorem will be used frequently with the hypothesis of constant mechanism effectiveness.

FUNDAMENTAL EFFICIENCY THEOREM
WITH CONSTANT MECHANISM EFFECTIVENESS

The mechanical efficiency η_m and cyclic shaft work W_s of an engine having a mechanism of constant effectiveness $0 < E \leq 1$ and constant buffer pressure are given by

$$\eta_m = E - \left(\frac{1}{E} - E\right)\frac{W_-}{W} \quad \text{and} \quad W_s = EW - \left(\frac{1}{E} - E\right)W_- \quad (3.2)$$

where W is the indicated work of the cycle and W_- is its forced work.

The theorem will be extended in Chapter 8 to include cases where buffer pressure varies and entails thermodynamic losses.

OPTIMUM BUFFER PRESSURE

If buffer pressure is varied relative to a fixed engine cycle, forced work W_- will vary along the lines illustrated in Figure 2.2. For a very low or for zero[†] buffer pressure, forced work will be present. As the buffer pressure level is increased, the forced work will decrease. Once entirely above the cycle, forced work will increase with no limit as buffer pressure is further increased. Because of the continuity assumptions of a regular cycle, at some intermediate point or points, forced work attains a minimum value. This is optimal for an engine with constant mechanism

---

[†] When buffer pressure is zero, the forced work of a regular cycle equals its absolute compression work (see Equation 1.2).

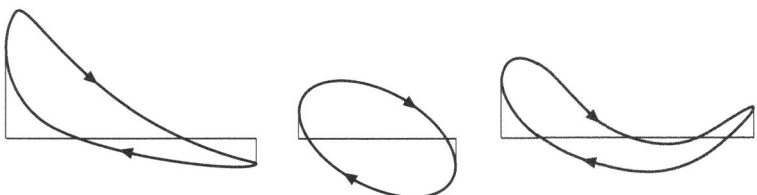

Figure 3.3 Examples of regular cycles together with their optimum buffer pressures.

effectiveness; since the indicated work W has remained fixed, Formulas (3.2) show this is a point of maximum mechanical efficiency and shaft work output. This proves the following result.

Every cyclic engine with a mechanism of constant effectiveness has an optimum constant buffer pressure level that maximizes mechanical efficiency.

Examples of non-efficacious regular cycles with optimum buffer pressures are given in Figure 3.3.

For an engine having a mechanism of variable effectiveness, it is not necessarily the case that maximum mechanical efficiency will occur where forced work is a minimum. However, the Fundamental Efficiency Theorem shows a sense in which constant effectiveness represents the best possible case, and knowing what influences the best case gives some intuition on how the subordinate cases are influenced. The closer a case is to the best possible, the more relevant becomes the knowledge of the best possible case.

OPTIMALLY BUFFERED STIRLING ENGINES

As shown above, the ideal Stirling with constant mechanism effectiveness represents the best case in a fair comparison class. This result is now carried a step further by considering optimal constant buffering of the ideal Stirling. Let τ represent the cold-to-hot *temperature ratio*

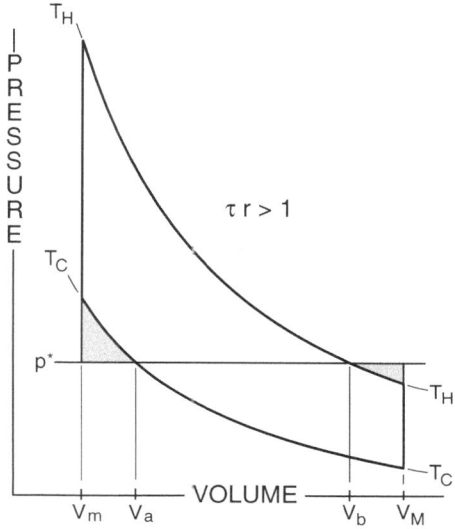

Figure 3.4 An non-efficacious ideal Stirling cycle with optimum buffer pressure.

of the two isotherms of the ideal Stirling cycle, and let r represent the ratio of the maximum to the minimum cycle volumes, which will be referred to as the *volume compression ratio*:

$$\tau = \frac{T_C}{T_H} \quad \text{and} \quad r = \frac{V_M}{V_m}.$$

The efficacious Stirling cycles are characterized by the inequality $\tau r \leq 1$. This ensures that the peak pressure on the lower isotherm T_C is not larger than the minimum pressure on the upper isotherm T_H. For any buffer pressure in between these, there is no forced work, as shown in Figure 1.8(d).

The case when $\tau r > 1$ is shown in Figure 3.4. The optimum buffer pressure p^* is the level where the total forced work, shown as the shaded area, is minimal. To see where this in fact occurs, imagine a buffer pressure set at any level intersecting both isotherms. If buffer pressure is increased from there, the forced work portion under the lower isotherm will decrease while that over the upper isotherm will increase. By the Fundamental Theorem of Calculus, the magnitude of the rate of change of each area equals the length of its segment on the buffer pressure line. One rate is positive (over the expansion isotherm) and the other is negative (under the compression isotherm). The net rate of area change is the algebraic sum of these. The minimum area will occur at the buffer level where this rate is zero. That point is exactly where the base lengths of the two areas shown in Figure 3.4 are equal; that is, the optimum buffer pressure p^* occurs where

$$V_a - V_m = V_M - V_b. \tag{3.3}$$

From this, p^* is easy to determine. On the lower and upper isotherms respectively, we have from the ideal gas law

$$p^* V_a = mRT_C \quad \text{and} \quad p^* V_b = mRT_H$$

where m is the mass of gas in the workspace and R is the ideal gas constant. Adding these and using the volume relation above gives

$$p^*(V_M + V_m) = mR(T_H + T_C)$$

which yields

$$p^* = \frac{mRT_H}{V_m} \frac{\tau + 1}{r + 1}. \tag{3.4}$$

Now expressions for V_a and V_b can be easily obtained and used in the calculation of the forced work areas:

$$
\begin{aligned}
W_- &= mRT_C \ln \frac{V_a}{V_m} - p^*(V_a - V_m) + p^*(V_M - V_b) - mRT_H \ln \frac{V_M}{V_b} \\
&= mRT_C \ln \frac{V_a}{V_m} - mRT_H \ln \frac{V_M}{V_b} = mRT_H\left[\tau \ln \frac{V_a}{V_m} - \ln \frac{V_M}{V_b}\right] \\
&= mRT_H[\tau \ln(mRT_C) - \tau \ln(p^* V_m) - \ln(p^* V_M) + \ln(mRT_H)] \\
&= mRT_H[\tau \ln \tau - (1 + \tau)[\ln(1 + \tau) - \ln(1 + r)] - \ln r].
\end{aligned}
$$

This formula is valid for the non-efficacious case. In the efficacious case the forced work is zero. Dividing by the indicated work of the ideal Stirling cycle, namely, $W = mRT_H (1 - \tau) \ln r$, the following formula is obtained.

THEOREM

The ratio of forced-to-indicated cyclic work in an optimally buffered ideal Stirling engine with temperature ratio τ and compression ratio r is given by

$$\frac{W_-}{W} = S(\tau, r)$$

where   (3.5)

$S(\tau, r) =$

$$\begin{cases} C & \text{if } \tau r \leq 1 \\ \dfrac{\tau \ln \tau - (1 + \tau)[\ln(1 + \tau) - \ln(1 + r)] - \ln r}{(1 - \tau) \ln r} & \text{if } \tau r > 1 \end{cases}$$

THE MECHANICAL EFFICIENCY LIMIT

The results above can be combined to yield a mathematically explicit upper bound on the mechanical efficiency of all engines in a fair comparison class regardless of buffer pressure levels.

Consider an arbitrary engine with compression ratio r, temperature extremes in the ratio τ, and a mechanism of effectiveness E or less. By the Stirling Comparison Theorem, its mechanical efficiency is less than or equal to that of an ideal Stirling engine with the same compression and temperature ratios, the same mass of working gas, a mechanism of constant effectiveness E, and the same buffer pressure. That mechanical efficiency in turn will be less than or equal to the same Stirling engine with the optimal buffer pressure. By the Fundamental Efficiency Theorem for constant mechanism effectiveness (3.2) and Formula (3.5) the following result is proven.

MECHANICAL EFFICIENCY LIMIT THEOREM

The mechanical efficiency of any cyclic engine with volume compression ratio r, effective temperature ratio τ, and mechanism effectiveness $\varepsilon \leq E$ cannot exceed

$$\eta_{ms}(E, \tau, r) = E - \left(\frac{1}{E} - E\right) S(\tau, r) \qquad (3.6)$$

where $S(\tau, r)$ is given by (3.5). This upper bound $\eta_{ms}(E, \tau, r)$ is attained by an ideal Stirling engine buffered at its optimal constant buffer pressure and having a mechanism of constant effectiveness E. Moreover, when $\tau r > 1$, it is the only engine with these E, τ, and r values to attain this efficiency.

This theorem reveals a limit on the mechanical efficiency of all kinematic engines with the same E, τ, and r values. The limit, $\eta_{ms}(E, \tau, r)$, is to mechanical efficiency what the well-known Carnot limit, $\eta_c = 1 - \tau$, is to thermal efficiency. It is an in-principle limit that no cyclic engine can exceed, and a limit that is attained by idealized Stirling engines.

An obvious use of the Mechanical Efficiency Limit Theorem is to provide quick measures of best possible efficiency. For example, any engine operating at a temperature ratio of $\tau = 0.5$, having a compression ratio of $r = 10$, and having a mechanism of effectiveness $E = 0.80$ or less cannot by (3.6) have a mechanical efficiency better than $\eta_{ms}(0.80, 0.5, 10) = 0.67$. The sensitivity of engine performance to changes in the basic parameters can also be easily ascertained. For example, if the present engine were to suffer a mechanical degradation in which the mechanism effectiveness dropped to $E = 0.60$ or less, then the mechanical efficiency of the engine would fall at least to $\eta_{ms}(0.60, 0.5, 10) = 0.29$, which is a practically fatal level of performance.

These are the mechancial efficiency limits. To obtain the overall brake thermal limits, one has only to multiply by the factor $(1 - \tau)$, as discussed in the next section.

THE BRAKE THERMAL EFFICIENCY LIMIT

The overall efficiency of an engine is measured by the ratio of shaft work obtained to the heat put in. It is usually referred to as the *brake thermal efficiency* and is simply the product of the thermal and mechanical efficiencies:

$$\eta_b = \frac{W_s}{Q_i} = \frac{W}{Q_i} \frac{W_s}{W} = \eta_t \, \eta_m.$$

The Carnot engine is not alone in having the maximum thermal efficiency. There is an entire class of cycles that potentially have the same efficiency. These are the so-called Reitlinger cycles, which consist of two isothermals and two polytropics of the same kind (Kolin, 1972). In these cycles, the heat absorbed on one polytropic is exactly equal to that rejected on the other, so regeneration is possible in principle. With perfect regeneration, the thermal efficiency of a Reitlinger cycle equals that of the Carnot.

Of all Reitlinger cycles, the Carnot is unique in requiring the least regeneration, namely, none at all, because its polytropics are adiabatics. However, its mechanical efficiency potential is inherently low. Figure 3.5 shows Carnot and Stirling cycles having the same

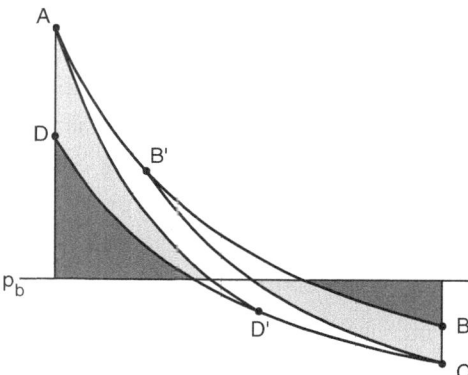

Figure 3.5 A Carnot cycle inscribed in a Stirling cycle with optimum constant buffer pressure. The forced work of the Stirling cycle ABCD is the darker shaded area; the forced work of the Carnot cycle AB'CD' also includes the lighter shaded area.

temperature and volume extremes. This example illustrates the typical higher forced work of the Carnot cycle.

The ideal Stirling is also a Reitlinger cycle. Its polytropics are isometric segments. The Stirling makes quite heavy demands on regeneration (Senft, 1993a), but with perfect regeneration[†] its thermal efficiency equals the Carnot. Because its mechanical efficiency is also maximal for fairly comparable engines, it has the maximum brake thermal efficiency potential.

BRAKE THERMAL EFFICIENCY COROLLARY

If η_b is the brake thermal efficiency of a cyclic heat engine with volume compression ratio r, temperature ratio τ, effectiveness $\varepsilon \leq E$, and constant buffer pressure, then

$$\eta_b \leq (1 - \tau)\, \eta_{ms}(E, \tau, r). \tag{3.7}$$

When $\tau r > 1$, equality holds if and only if the engine is an ideal Stirling with perfect regeneration and a mechanism of constant effectiveness E.

AVERAGE CYCLE AND OPTIMUM BUFFER PRESSURE

The optimum buffer pressure for the non-efficacious Stirling cycle is not identical to the mean cycle pressure, but it is typically close. This is fortunate because most practical Stirling engines operate with their mean pressure equal to the buffer pressure. This is because there is always some unavoidable leakage past piston or rod seals in a real engine. Without special pumps or check valves, or seals that function as such, the average pressure in the workspace comes to more or less equal the external pressure once steady state operation is attained.

[†] In practice, it is relatively easy to make good regenerators for a Stirling engine.

The following definitions have been proposed (Senft, 1982) for any regular cycle:

$$p_{mu} = W_e/(V_M - V_m) = \textit{mean upper pressure}$$
$$p_{ml} = W_c/(V_M - V_m) = \textit{mean lower pressure} \qquad (3.8)$$
$$p_m = (p_{mu} + p_{ml})/2 = \textit{mean cycle pressure.}$$

Applying definitions (3.8) to the ideal Stirling cycle yields

$$p_m = \frac{mRT_H(1 + \tau)\ln r}{2V_m(r-1)}. \qquad (3.9)$$

Therefore, from (3.4),

$$\frac{p_m}{p^*} = \frac{(1+r)\ln r}{2(r-1)}. \qquad (3.10)$$

It is interesting that this ratio does not depend upon the cycle temperatures, but only upon the compression ratio. One can easily prove that it is always true that $p_m > p^*$ because $r > 1$, but calculations for a range of compression ratios used on real Stirling engines show the two pressures are quite close, as Table 3.1 illustrates.

Table 3.1 *Comparison of mean cycle and optimum buffer pressures for ideal Stirling cycles with various compression ratios.*

r	p_m/p^*
1.1	1.001
1.2	1.003
1.3	1.006
1.5	1.014
2.0	1.040
3.0	1.099

4

COMPRESSION RATIO
AND SHAFT WORK

Derived in this chapter are two direct consequences of the Mechanical Efficiency Limit Theorem proven in the preceding chapter. One consequence is a surprising intrinsic limitation on the geometry of certain engines. The other consequence is the determination of the maximum shaft work that can be obtained from engines under common restrictions. This yields a theoretical standard to which particular engines can be conveniently compared.

LIMITS ON COMPRESSION RATIO

That there are restrictions on the compression ratio one can use in an engine is an unexpected implication of the Mechanical Efficiency Limit Theorem. This limitation can be seen by considering the behavior of $\eta_{ms}(E, \tau, r)$ as a function of r for fixed E, and τ. Maximal mechanical efficiency η_{ms} is simply equal to E as long as $r \leq 1/\tau$. For $r > 1/\tau$, it is easily determined by elementary calculus that $\eta_{ms}(E, \tau, r)$ is monotone decreasing with increasing r. It is also readily found that $\lim_{r \to \infty} S(\tau, r) = \tau/(1 - \tau)$, which gives

$$\lim_{r \to \infty} \eta_{ms}(E, \tau, r) = \frac{E^2 - \tau}{E(1 - \tau)}. \tag{4.1}$$

Thus, if $E^2 > \tau$, η_{ms} will asymptotically approach a positive lower bound. However, when $E^2 < \tau$, then limit (4.1) is negative. This means η_{ms} decreases and becomes zero at a finite value of r. When its mechanical efficiency is zero (or less!), an engine cannot run. By the Mechanical Efficiency Limit Theorem, no engine can have a better mechanical efficiency than η_{ms}, so no engine can run with a compression ratio beyond the r value where η_{ms} becomes zero. This proves the following result.

COMPRESSION RATIO LIMIT THEOREM

For a given E and τ with $E^2 < \tau$, there is a compression ratio value above which no cyclic engine with mechanism effectiveness $\varepsilon \leq E$ and temperature ratio τ will operate. This is the r value where $\eta_{ms}(E, \tau, r) = 0$.

Figure 4.1 shows graphs of maximal mechanical efficiency η_{ms} with respect to r for specific values of E and τ. The smaller E^2 is relative to τ, the faster η_{ms} becomes zero. This means that for engines operating from relatively low temperature heat sources, such as waste industrial process heat or passive solar energy, the range of usable compression ratio is definitely limited. The closer τ is to 1, the more limited the compression ratio range becomes. This is precisely why low temperature differential engines require low compression ratios (Senft, 1987b).

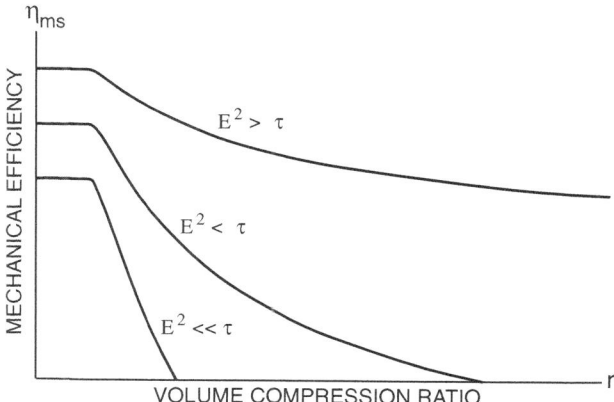

Figure 4.1 Graphs of maximal mechanical efficiency with respect to compression ratio shown for three different values of mechanism effectiveness E and fixed temperature ratio τ.

To propose an engine with, say, a ten-to-one compression ratio to operate from passive solar energy is to not be in possession of this fundamental understanding.

On the other hand, if an engine is to operate from a very high temperature source, as is the case, for example, with internal combustion engines, then $\tau = T_C/T_H$ will be small enough so that $E^2 > \tau$ for all practical values of E. In this case, there is no intrinsic restriction on compression ratio, and it can be chosen to suit any other requirements or desires. Now, mechanical efficiency does decrease with increasing compression ratio even in this case, but the limit equation (4.1) gives an easily calculated lower bound. For example, if $\tau = 0.3$ and $E = 0.9$, then no matter how large a compression ratio is chosen, η_{ms} would still be at least 0.81:

$$\eta_{ms}(0.9, 0.3, r) \geq \frac{(0.9)^2 - 0.3}{0.9(1 - 0.3)} \doteq 0.81.$$

SHAFT WORK LIMITS

The preceding results might tempt one to conclude that all engines should be made efficacious, that is, with $r \leq 1/\tau$. Then the mechanical efficiency bound would agree with the mechanism effectiveness bound, which is the best one can hope for. This is usually the case with high-speed Stirling engines having high temperature heat sources. Widely separated temperatures give a small temperature ratio (e.g., $\tau \sim 0.3$), and high speeds require relatively large internal heat exchanger void volumes, which produce lower compression ratios (e.g., $r \leq 2$). So these engines turn out efficacious ($\tau r \leq 1$). However, engines operating from lower temperature heat sources do not have to be made efficacious. As will be seen in this section, non-efficacious engines offer advantages when size relative to power output is an important objective.

The Mechanical Efficiency Limit Theorem gives an upper bound on obtainable cyclic shaft work by taking the product of η_{ms} with the *indicated cyclic work of the ideal Stirling cycle*:

$$W = m R \left(T_H - T_C\right) \ln r. \tag{4.2}$$

The obvious upper bound on cyclic shaft work for all possible engines would thus be $\eta_{ms}(E, \tau, r) m R \left(T_H - T_C\right) \ln r$. Although this indeed is an upper bound, it is not as useful as it might at first seem because of its explicit dependence upon m. In practice it is usually not known how much gas m is taking part in the cycle. Moreover, the mass content will change through net leakage into or out of the workspace if, say, operating temperatures or the compression ratio is changed.

It is more convenient and more certain to work instead with the mean pressure of the working gas as a basis for evaluation. As already mentioned, the mean pressure of the working gas usually adjusts through leakage to more or less equal the external or buffer pressure, which is easily measured. In cases where the workspace mean pressure is elevated above the buffer pressure by a check valve or a pump, mean workspace pressure can be monitored with a damped pressure gauge connected through a capillary tap into the engine capsule.

With mean cycle pressure as a given parameter, a parameter capturing the size of the engine is also required for making meaningful comparisons. Specifiying only swept volume would not yield fair comparisons because that would allow very large engines to be compared to very small ones. Rather, the total or maximum workspace volume V_M pretty well characterizes the overall size of an engine. The term *mean-pressure total-volume specific cyclic work*, or just *specific work* for short, will be used to mean cyclic work divided by mean pressure and maximum workspace volume. Thus, the *indicated specific cyclic work* $\hat{W}$ of an engine is

$$\hat{W} = \frac{W}{p_m V_M}.$$

Expression (4.2) and Formula (3.9) yield an explicit formula for the *specific indicated work of an ideal Stirling engine*, namely

$$\frac{2(r - 1)(1 - \tau)}{r(1 + \tau)} \tag{4.3}$$

The product of $\eta_{ms}(E, \tau, r)$ with (4.3) gives the corresponding specific shaft work of an ideal Stirling engine. That this is the maximum obtainable among comparable engines is stated precisely in the following theorem.

MAXIMUM SHAFT WORK THEOREM

The specific cyclic shaft work

$$\hat{W}_s = \frac{W_s}{p_m V_M}$$

of any engine having compression ratio r, temperature extremes in the ratio τ, and a mechanism of effectiveness E or less is bounded above by

$$\hat{W}_s^* = 2 \frac{(r-1)(1-\tau)}{r(1+\tau)} \eta_{ms}(E, \tau, r). \tag{4.4}$$

Thus, the per cycle shaft work of any engine cannot exceed the product of its mean pressure, maximum workspace volume, and $\hat{W}_s^*$ given by (4.4). This is an easy and practical way to estimate the best shaft output an engine can give. The closer the engine cycle is to the ideal Stirling, of course, the closer will be its output to this upper bound. A proof of this theorem, different than that first presented (Senft, 1993b), is given in the last section of this chapter. The theorem is also valuable for the general insights it implies, some of which will now be described.

Figure 4.2 shows graphs of the maximal specific shaft work $\hat{W}_s^*$ versus compression ratio for three different sets of values for E and τ. A point of great interest is that there are maxima, and that these can occur in the non-efficacious range, Figure 4.3 illustrates this in more detail. The subject of optimum compression ratio is examined further in Chapter 7.

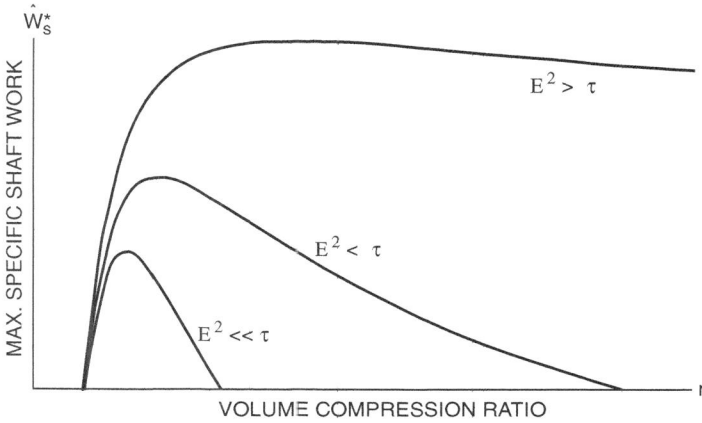

Figure 4.2 Graphs of maximum shaft work for given mean pressure and total volume versus compression ratio.

TEMPERATURE EFFECTS

The results above also clearly explain the behavior of engines operating, or trying to operate, between a source and sink having a small temperature differential. Figure 4.3 is another view of maximal specific shaft work $\hat{W}_s^*$ given by (4.4) versus compression ratio r, but for fixed mechanism effectiveness of $E = 0.75$ and temperature ratio τ varying from 0.75 up to 0.90. If the sink temperature T_C is fixed, these curves correspond respectively to decreasing source temperatures T_H.

Now, these graphs represent specific shaft output of ideal Stirlings and so represent the maximum possible shaft output for all engines with the same E, τ, and r values, and the same mean pressure and maximum workspace volume. Thus, the output curves for any particular (non ideal Stirling) engine will be contained under one of these. The array of curves for other values of E is similar, as the previous figure indicates. Thus is confirmed the observation made earlier that *for a fixed mechanism effectiveness, engines operating*

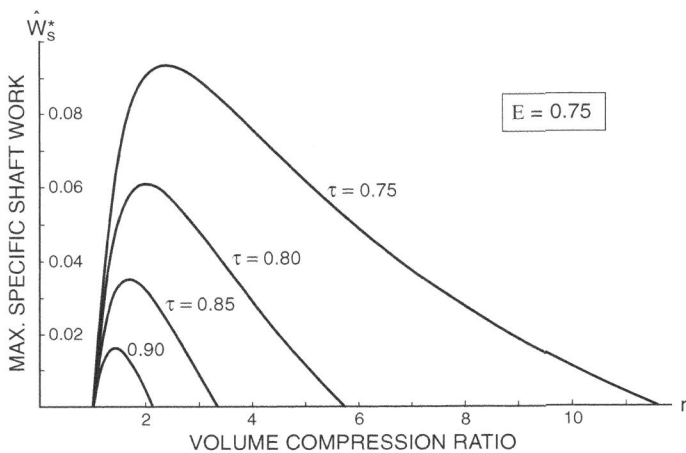

Figure 4.3 Graphs of maximum specific shaft work versus compression ratio for engines having a mechanism effectiveness E = 0.75 operating with various temperature ratios.

between smaller temperature differences require lower compression ratios than engines operating between larger temperature differences (Senft, 1987b).

The existence of the optimum compression ratios shown in Figure 4.3 also illustrates a fact that at first appears to be counterintuitive: Sometimes a decrease in compression ratio can increase engine output, and vice versa. This is obvious once Figure 4.3 is seen, of course. The output of an ideal Stirling operating with a temperature ratio of 0.80 follows the second largest curve in the figure. With a compression ratio of 5, the specific output is about 0.01. If the compression ratio is decreased to 4, output increases to almost 0.03. The maximum output of about 0.06 is realized with a compression ratio of a little over 2. Note that this optimum occurs well within the non-efficacious range: $\tau r = (0.80)(2) = 1.6 > 1$. In fact, at the efficacious boundary point, $r = 1/\tau = 1/0.80 = 1.25$, the specific output is down to 0.033, about half the maximum output.

The behaviors described here are for an ideal Stirling engine, which practical engines will not match. But one can expect the same kind of trends in practical engines since the curves given here represent universal maximums. The more the cycle characteristics are like the ideal Stirling, the closer will be the behavior. This has been observed in the laboratory on small low-temperature differential Stirling (Kolin, 1983, p. 119) and Ringbom (Senft, 1986) engines. Non ideal cycles may differ in details; for example, the optimum compression ratios may be at or even below the efficacious point as Chapter 7 shows. Nevertheless, the ideal the maximum shaft work curves impose certain inescapable constraints inasmuch as they overspread and incase the curves of all engines within a fair comparison class.

The intuition behind the counterintuitive effects of changing compression ratio can be seen in the following way. Figure 4.4 shows a group of ideal Stirling cycles with the same mean cycle pressure and equal buffer pressure. The first three cycles have the same volume extremes and the same sink temperature T_C. The first cycle has a source temperature sufficiently high that there is no associated forced work. Assuming the best possible case here of constant mechanism effectiveness E, the shaft output is simply E multiplied by the indicated work.

In the second cycle, the source temperature is lower. Some forced work appears, diminishing mechanical efficiency according to the Fundamental Efficiency Theorem (3.2):

$$\eta_m = E - (1/E - E)(W_-/W).$$

Note that one might say the reduction is of the second order because indicated work decreases also.

In the third cycle, the source temperature is still lower; there is much more forced work and less indicated work. At this point, the engine is producing much less at the shaft than before. Further decreases in the source temperature would, of course, make things

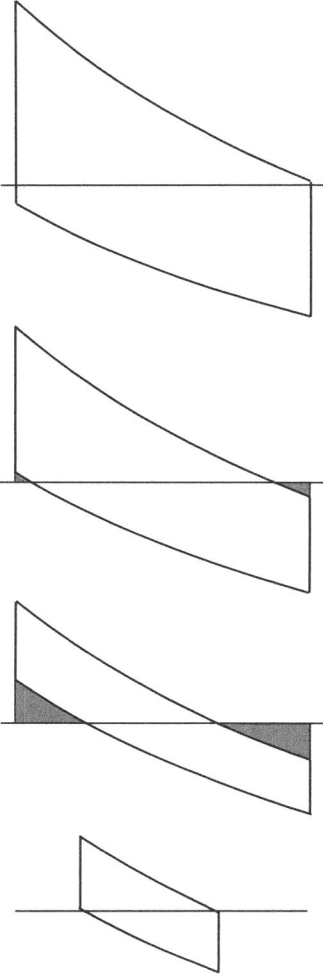

Figure 4.4 A sequence of ideal Stirling cycles illustrating the effects of decreasing source temperature.

worse and would eventually reach the point where mechanical efficiency becomes zero and the engine is unable to run.

The last cycle shows what effect reducing the compression ratio would have. Forced work can be made to vanish, mechanical efficiency becomes positive again (it becomes E, in fact), and the engine can run again! This phenomenon can be readily observed in model Stirling engines, especially the low-temperature differential variety. In fact, it was this observation, and attempts to explain it, that led to the Fundamental Efficiency Theorem (Senft, 1985, 1987a).

PROOF OF THE MAXIMUM SHAFT WORK THEOREM

In view of the Mechanical Efficiency Limit Theorem (3.6), it suffices to show that Expression (4.3) is an upper bound on specific indicated work. Suppose an arbitrary engine A is under consideration with maximum volume V_M, mean pressure p_m, and compression ratio r. Using Definitions (3.8),

$$p_m = \frac{W_e + W_c}{2(V_M - V_m)}$$

and thus the specific indicated work of engine A can be expressed as

$$\hat{W} = \frac{W}{p_m V_M} = \frac{(W_e - W_c) 2 (V_M - V_m)}{(W_e + W_c) V_M} = 2 \frac{r-1}{r} \frac{W_e - W_c}{W_e + W_c}. \quad (4.5)$$

Now let τ denote the ratio of the extreme temperatures achieved by the cycle A. Select for comparison an ideal Stirling cycle Ω having the same temperature and volume extremes, and the same mass of the same ideal gas as engine A. Then the cycle A will be inscribed in Ω as shown in Figure 4.5 in the p–V plane.

Because the expansion process of A is below that of Ω,

$$W_e \leq mRT_H \ln r.$$

Because the compression process of A is above that of Ω,

$$W_c \geq mRT_C \ln r.$$

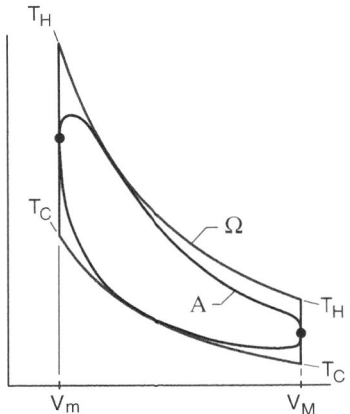

Figure 4.5 An arbitrary engine cycle A inscribed in an ideal Stirling cycle Ω.

Therefore, the ratio of absolute compression to expansion work of A satisfies

$$\frac{W_c}{W_e} \geq \frac{T_C}{T_H} = \tau \,.$$

Hence,

$$\frac{W_e - W_c}{W_e + W_c} = \frac{1 - (W_c/W_e)}{1 + (W_c/W_e)} \leq \frac{1 - \tau}{1 + \tau} \,.$$

Combining this with (4.5) establishes what was to be shown.

5

PRESSURIZATION EFFECTS

Formula (4.2) for the indicated cyclic work of an ideal Stirling engine immediately suggests that output can be increased by *charging* the workspace with more working gas, keeping everything else the same. This is the motivation behind *pressurizing* or supercharging an engine. What matters in the end, of course, is whether shaft output improves, and this is a matter of mechanical efficiency.

An easy case to understand at this point is that of an ideal Stirling engine having a constant mechanism effectiveness and optimum buffer pressure. Its mean workspace pressure would be proportional to m, as Formula (3.9) explicitly shows. The Maximum Shaft Work Theorem (4.4) thus implies that if the engine has the charge of its working gas increased by a certain factor, and its buffer pressure adjusted to be optimal for the new charge (in fact, it will need to be increased by exactly the same factor, as Formula (3.4) shows), the shaft output will increase by the same factor. Hence, pressurizing an optimal ideal Stirling in this way will increase output in direct proportion to the charge factor. This kind of pressurization, called *system charging*, where the workspace and buffer pressure are charged together uniformly by the same factor, produces the same best possible results in many engine and buffer pressure combinations.

SYSTEM CHARGING MONOMORPHIC ENGINES

A *monomorphic* engine is one in which its pressure–volume function is proportional to the gas mass content of its workspace. For example, ideal Stirling cycles have expansion and compression pressure functions $p_e(V) = mRT_H/V$ and $p_c(V) = mRT_C/V$, respectively, so with fixed temperature and volume extremes both are proportional to m. The same is true for the Otto cycle, and indeed for all Crossley cycles (Kolin, 1972). In the case of the Carnot cycle, each of its four pressure functions are proportional to m; this is true more generally for all Reitlinger cycles. All these cycles and many other nameless classes of cycles with fixed temperature and volume extremes are monomorphic.

It is convenient to represent engine cycles in the p–V plane parametrically. Mathematically, an engine cycle can be defined as a continuous piecewise smooth closed curve in the p–V plane, that is, a closed curve which is the union of a finite number of endpoint-connected smooth curves. The parametric description can be simply given as

$$\Gamma = (V, p) = (\alpha(s), \beta(s)), \qquad s \in J = [a, b]$$

where the functions α and β are positive on J, which is a closed bounded interval of length equal to the least common period of the two functions. The indicated work of the engine is then succinctly expressible as

$$W\langle\Gamma\rangle = \int_\Gamma p \, dV = \int_a^b \beta(s)\,\alpha'(s)\,ds \qquad (5.1)$$

The efficacious and forced works of the engine can be easily expressed by making use of the standard positive and negative part functions (Rudin, 1964). These part functions are defined as

$$z^+ = \max\{0, z\} \quad \text{and} \quad z^- = \max\{0, -z\}.$$

Both part functions are non-negative by definition, and the following identities hold:

$$\begin{aligned} z &= z^+ - z^- \\ (xy)^- &= x^+ y^- + x^- y^+ \\ (kz)^- &= k z^- \quad \text{if} \quad k > 0. \end{aligned} \qquad (5.2)$$

The forced work of the cycle $\Gamma = (V, p)$ relative to the buffer pressure p_b can then be written as

$$W_-\langle\Gamma\rangle = \int_\Gamma [(p - p_b)\,dV]^- . \tag{5.3}$$

By the third identity of (5.2), this is calculated as

$$\int_a^b [(\beta(s) - p_b)\,\alpha'(s)]^-\,ds$$

since $ds > 0$ in the elementary integral. Using the second identity of (5.2), the forced work can also be written as

$$W_- = \int_\Gamma (p - p_b)^+ (dV)^- + \int_\Gamma (p - p_b)^- (dV)^+$$

which clearly shows the two components that make up forced work. One component occurs during the parts of compression $dV < 0$ where $p > p_b$; the other component comes from expansion $dV > 0$ whenever $p < p_b$.

Suppose now that the charge of the engine Γ is increased by the factor k. If the engine is monomorphic, then the new cycle can be expressed as $\Gamma_k = (V, kp)$; that is, the new pressure function is $kp = k\beta(s)$ and the volume function $V = \alpha(s)$ is unchanged. Therefore, by (5.1), $W\langle\Gamma_k\rangle = kW\langle\Gamma\rangle$. If the buffer pressure for the cycle Γ_k is also increased by the factor k, then, using the third identity of (5.2), the forced works of the two engines also stand in the ratio k:

$$W_-\langle\Gamma_k\rangle = \int_a^b [(k\beta(s) - kp_b)\,\alpha'(s)]^-\,ds$$

$$= k \int_a^b [(\beta(s) - p_b)\,\alpha'(s)]^-\,ds = kW_-\langle\Gamma\rangle.$$

For an engine with constant mechanism effectiveness E, the Fundamental Efficiency Theorem (3.2) shows its shaft work is also increased by the factor k:

$$W_s\langle\Gamma_k\rangle = EW\langle\Gamma_k\rangle - \left(\frac{1}{E} - E\right)W_-\langle\Gamma_k\rangle$$

$$= EkW\langle\Gamma\rangle - \left(\frac{1}{E} - E\right)kW_-\langle\Gamma\rangle$$

$$= k\left[E\,W\langle\Gamma\rangle - \left(\frac{1}{E} - E\right)W_-\langle\Gamma\rangle\right] = kW_s\langle\Gamma\rangle.$$

This proves the following theorem.

SYSTEM CHARGING THEOREM

In any monomorphic engine with constant mechanism effectiveness, if the charge of its workspace and its buffer pressure are increased by the same factor, then its shaft work also will increase by the same factor, and mechanical efficiency is preserved.

This result is also true in the more general setting of engines having nonconstant buffer pressure (Senft, 1991b).

ENGINES CHARGED ABOVE BUFFER PRESSURE

An engine in which the buffer pressure never exceeds the workspace pressure will be referred to as being *charged above buffer pressure* or as *buffered from below*. There are a number of practical advantages to buffering from below. These advantages include preventing lubricant migration into the workspace, preventing or minimizing bearing load reversals, and perhaps most important, increasing output, albeit at the expense of mechanical efficiency. It often is the case in practice that buffer pressure cannot be modified, as in engines that have a crankcase open to the atmosphere or as in the old hot air water pumping engines, which have no crankcase at all. Charging the workspace alone is the only option to increase the output of such engines. Even when a crankcase is totally enclosed and pressureworthy, the other advantages may make charging above buffer pressure desirable, as was the case with the Philips 102C (Hargreaves, 1991) and Ross (1987) engines.

For engines buffered from below, there is no forced work arising during expansion. The forced work occurs over the whole compression process and is simply the absolute compression work (1.2) minus the area below the buffer pressure line:

$$W_- = W_c - p_b \, \Delta V.$$

This is illustrated in Figure 5.1 for a generic regular cycle.

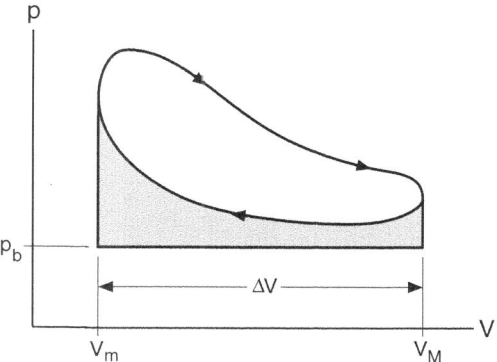

Figure 5.1 A regular cycle buffered from below. The shaded area is the forced work associated with the cycle and buffer pressure.

Immediate from the Fundamental Efficiency Theorem of Chapter 3, is the following theorem.

THEOREM

If an engine with a regular cycle having a mechanism of constant effectiveness $0 < E \leq 1$ is buffered from below by constant pressure p_b then its shaft work is given by

$$W_s = E\,W - \left(\frac{1}{E} - E\right)(W_c - p_b\,\Delta V) \qquad (5.4)$$

where $\Delta V = V_M - V_m$ is the volume variation of the engine workspace, W_c is the absolute compression work of the cycle, and W is its indicated work.

This theorem has interesting consequences concerning the minimal conditions under which an engine is able to operate. To say an engine can run means that it produces useful shaft work or, at the very least, just manages to sustain its own operation, namely, $W_s \geq 0$. By the theorem above, this inequality yields a lower bound on E in terms of p_b and the cycle characteristics.

THEOREM

An engine buffered from below will run if and only if

$$E^2 \geq \frac{W_c - p_b \Delta V}{W_e - p_b \Delta V}.$$

The hypotheses here are the same as above, and W_e is the absolute expansion work of the cycle. Reducing the fractional expression in the theorem by ΔV, the condition can be stated in an interesting way using pressures only:

$$E^2 \geq \frac{p_{ml} - p_b}{p_{mu} - p_b}$$

where p_{mu} and p_{ml} are the mean upper and lower pressures of the cycle as defined in (3.8).

From elementary algebra,

$$\frac{a}{b} \geq \frac{a - c}{b - c} \quad \text{whenever} \quad 0 \leq c \leq a < b$$

which yields the following consequence of the previous theorem:

COROLLARY

If an engine satisfies $E^2 \geq W_c/W_e$ then it will run when buffered from below at any pressure (including $p_b = 0$).

In the case of ideal Stirlings, the absolute work ratio is simply the temperature ratio, that is, $W_c/W_e = T_C/T_H = \tau$. Another note of interest is that Obert (1960) defined a concept called "work ratio" for cycles as $r_w = W/W_e$ when expressed in the nomenclature of this book. It is relevant in this context only when buffer pressure is zero; then it can be used in the statement of the preceding corollary as

$$E^2 \geq \frac{W_c}{W_e} = 1 - \frac{W}{W_e} = 1 - r_w$$

THE WORKSPACE CHARGING THEOREM

In any engine buffered from below and having a mechanism of constant effectiveness, the shaft work formula (5.4) can be rearrranged to

$$W_s = W_c \left(\mathrm{E} \frac{W_e}{W_c} - \frac{1}{\mathrm{E}} \right) + \left(\frac{1}{\mathrm{E}} - \mathrm{E} \right) p_b \, \Delta V.$$

Consider charging only the workspace and leaving the buffer pressure fixed. Then the second term is constant for a given engine. In a monomorphic engine, W_c / W_e is constant, so the first term is directly proportional to m. Thus, shaft work is a linear function of m for a monomorphic engine when buffered from below. The sign of its slope is the sign of the factor

$$\mathrm{E} \frac{W_e}{W_c} - \frac{1}{\mathrm{E}}.$$

If this is positive, then shaft work increases as the workspace is charged. If it is negative, then shaft work decreases to and becomes zero at a finite charge level, and the engine stalls. Furthermore, dividing the shaft work expression above through by indicated work W gives the following equation for the mechanical efficiency of the engine:

$$\eta_m = \frac{W_c}{W} \left(\mathrm{E} \frac{W_e}{W_c} - \frac{1}{\mathrm{E}} \right) + \left(\frac{1}{\mathrm{E}} - \mathrm{E} \right) \frac{p_b \Delta V}{W}. \tag{5.5}$$

Now W is proportional to m for a monomorphic engine, and therefore the first term on the right is constant with respect to m while the second decreases to zero as $m \to \infty$. These observations constitute a proof of the Workspace Charging Theorem.

WORKSPACE CHARGING THEOREM

Suppose a given monomorphic engine with constant mechanism effectiveness E is buffered by a constant pressure. Then the following statements hold for values of the charge mass m for which the engine is buffered from below:

- If $\mathrm{E}^2 > W_c / W_e$ then W_s is a positive increasing function of m.
- If $\mathrm{E}^2 < W_c / W_e$ then W_s is a decreasing function of m, and $W_s = 0$ at a finite value of m.
- In either case, η_m is a decreasing function of m.

The theorem has interesting implications. In the first case, namely if $E^2 > W_c/W_e$, the useful work output of the engine can be increased by pumping up the workspace, and any desired level of output can be obtained by charging high enough. In the second case, if $E^2 < W_c/W_e$, this simple trick will not work and in fact will make matters worse: if carried too far it is guaranteed to render the engine lifeless. This part could appropriately be called the *Stall Theorem*. The third item could be called the *Theorem of Diminishing Returns* because it guarantees that even in the best case, where output can be increased by charging, it can be done only at the expense of diminished mechanical efficiency. There is, however, a positive lower bound in this case, namely, the first term on the right-hand side of (5.5); this can be written as

$$\eta_m \geq \frac{E^2 - (W_c/W_e)}{E(1 - W_c/W_e)}. \tag{5.6}$$

6

CHARGE EFFECTS
IN IDEAL STIRLING ENGINES

Ideal Stirling engines are monomorphic, so the results of the previous chapter apply directly. System charging an ideal Stirling theoretically produces the best possible effect. This is so whether the engine is optimally buffered or not; output increases and mechanical efficiency remains constant.

The Workspace Charging Theorem also applies. Its criteria take a simple form because, for an ideal Stirling, the absolute work ratio equals the temperature ratio:

$$W_c/W_e = T_C/T_H = \tau.$$

Thus, if $E^2 > \tau$, cyclic shaft work output will increase indefinitely as the workspace is charged higher and higher above a fixed buffer pressure. Of course, mechanical efficiency will certainly decrease after the engine becomes buffered entirely from below.

If, on the other hand, $E^2 < \tau$, then shaft output decreases as the workspace is charged above buffer pressure, and eventually the engine will even be unable to run itself. This is just what the Workspace Charging Theorem tells us. Still a question in this case is how output varies in the range just prior to the point where the engine becomes charged above the buffer pressure. This is an important question if one is seeking to improve the performance of an engine by adjusting only its workspace charge level. This question is answered in detail in this chapter for ideal Stirling engines.

WORKSPACE CHARGING IDEAL STIRLING ENGINES

Assumed throughout this analysis will be a mechanism of constant effectiveness E and fixed constant buffer pressure p_b. The formula for the cyclic shaft work output from the Fundamental Efficiency Theorem under this assumption is

$$W_s = EW - \left(\frac{1}{E} - E\right) W_- . \tag{3.2}$$

Let m_o denote the gas mass for which the Stirling cycle is just barely buffered from above, that is, for which the peak engine workspace pressure equals the buffer pressure. This is shown in the first cycle of Figure 6.1. This is not at all a desirable cycle and buffer pressure combination in practice, but it is a mathematically natural reference point. The gas mass content at this point is given by

$$m_o = \frac{p_b V_m}{RT_H} . \tag{6.1}$$

In the present analysis it is convenient to work with the gas mass normalized by this m_o value; thus, a dimensionless gas mass variable M is defined as

$$M = m/m_o . \tag{6.2}$$

It is also convenient to use dimensionless work quantities. In view of (6.1), it is appropriate to normalize relative to the fixed buffer pressure p_b and minimum cycle volume V_m. Combining (6.1), (6.2), and (4.2) gives the following simple expression for the *normalized indicated cyclic work*

$$\overline{W} = W/(p_b V_m) = M(1 - \tau) \ln r . \tag{6.3}$$

The overline will be used throughout this analysis to denote work quantities which are normalized by dividing by $p_b V_m$. The normalized forced work function $\overline{W_-}$ requires some care to express as a function of M, because there are two types of engine cycles to consider and five cases for each of these.

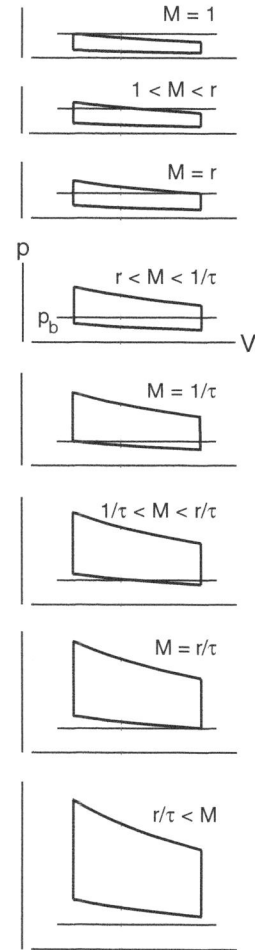

Figure 6.1 Pressure–volume diagrams of an efficacious Stirling cycle shown at various workspace charge levels M against a fixed buffer pressure. For these cycles, the temperature ratio $\tau = 1/3$, and the compression ratio $r = 3/2$.

Efficacious Cycles

To calculate the forced work of a Stirling cycle for all possible work-space charge levels against a fixed buffer pressure, five intervals must be considered. The intervals arise from the four gas mass values that put the corners of the cycle on the buffer pressure line.

If $\tau r \leq 1$, the Stirling cycle has the property that the minimum expansion process pressure is greater than or equal to the maximum compression process pressure. For these so-called efficacious Stirlings, the four points expressed in terms of the dimensionless mass variable M are

$$1 < r \leq \frac{1}{\tau} < \frac{r}{\tau}. \tag{6.4}$$

Figure 6.1 shows an example of an efficacious Stirling cycle with eight charge mass M levels and fixed buffer pressure p_b. The work-space charge progresses from $M = 1$, where the cycle is just below the buffer pressure, to $M > r/\tau$, where the cycle is completely above the buffer pressure. A straightforward series of calculations yields the following expression for the *normalized forced work* $\overline{W_-} = W_-/(p_b V_m)$ *for an efficacious Stirling*:

$$\overline{W_-}(M) = \begin{cases} r - 1 - M \ln r & \text{if } 0 \leq M \leq 1; \\ r - M(1 - \ln M + \ln r) & \text{if } 1 \leq M \leq r; \\ 0 & \text{if } r \leq M \leq 1/\tau; \\ M\tau(\ln M + \ln \tau - 1) + 1 & \text{if } 1/\tau \leq M \leq r/\tau; \\ M\tau \ln r - r + 1 & \text{if } r/\tau \leq M. \end{cases}$$

Using this and (6.3) for $\overline{W}$ in (3.2) gives the formula for *normalized shaft work*:

$$\overline{W_s}(M) = E \overline{W}(M) - \left(\frac{1}{E} - E\right) \overline{W_-}(M).$$

Figure 6.2 shows graphs of $\overline{W_s}(M)$ for an efficacious engine with $r = 3/2$ and $\tau = 1/3$. This temperature ratio is typical of high performance Stirling engines with operating temperatures in the neighborhood of $T_H = 700°C$ and $T_C = 50°C$. Curves are shown in the figure

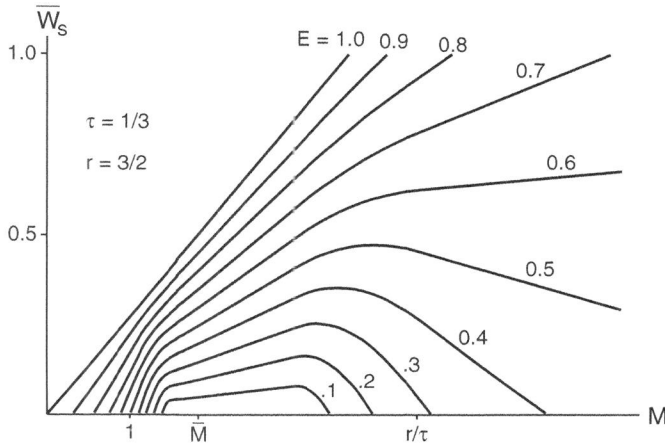

Figure 6.2 Normalized cyclic shaft work $\overline{W}_s(M)$ versus charge mass factor M for an effi-cacious ideal Stirling engine having temperature ratio $\tau = 1/3$ and compression ratio $r = 3/2$. Graphs are shown for constant mechanism effectiveness values E ranging from 1.0 down to 0.1.

for mechanism effectiveness values ranging from $E = 1$ (no friction loss) down to a miserably low $E = 0.1$. Values of E less than about 0.6 really do not occur in well-designed, well-built, and properly main-tained engines, but low values are included in the graphs here to show the theoretical effects of poor mechanism performance.

Marked on the M-axis in the figure are two significant points. The point

$$\overline{M} = \frac{2(r-1)}{(\tau+1)\ln r}$$

marks the charge level factor where the average pressure of the engine cycle p_m given by (3.9) equals the fixed external pressure p_b. The other significant point is r/τ which is the charge level factor where the engine is just barely buffered from below, that is, where the lowest corner of the cycle touches the buffer pressure line. This is shown in the next to lowest cycle in Figure 6.1. For higher charge levels, the cycle goes com-pletely above the buffer pressure line. The characteristics of the shaft work graphs for all efficacious Stirlings are the same as those exhibited in Figure 6.2 (Senft, 1989).

Non-Efficacious Cycles

Non-efficacious Stirling cycles are those for which $\tau r > 1$. The interval decomposition needed to calculate the forced work in this case is slightly different than (6.4). This is because the minimum expansion pressure is below the maximum compression pressure. Again the interval endpoints correspond to the gas charges which put the workspace cycle corners on the buffer pressure line:

$$1 < \frac{1}{\tau} < r < \frac{r}{\tau}.$$

Figure 6.3 shows a non-efficacious cycle with eight charge levels ranging from below to above buffer pressure. The *normalized forced work for non-efficacious* Stirling cycles is given by the following formulas:

$$\overline{W_-}(M) = \begin{cases} r - 1 - M \ln r & \text{if } 0 \leq M \leq 1; \\ r - M(1 - \ln M + \ln r) & \text{if } 1 \leq M \leq 1/\tau; \\ r + M[(\tau + 1)(\ln M - 1) \\ \quad + \tau \ln \tau - \ln r] + 1 & \text{if } 1/\tau \leq M \leq r; \\ M\tau(\ln M + \ln \tau - 1) + 1 & \text{if } r \leq M \leq r/\tau; \\ M\tau \ln r - r + 1 & \text{if } r/\tau \leq M. \end{cases}$$

Note that the formulas for forced work in the two cases differ only in the middle interval and also that the adjacent intervals have different endpoints. Figure 6.4 shows graphs of normalized shaft work $\overline{W_s}$ for the non-efficacious engine with $\tau = 1/2$ and $r = 5/2$. This family of graphs is representative of all non-efficacious ideal Stirling engines.

PRACTICAL IMPLICATIONS

From the above analysis, the effect on shaft work of charging only the workspace is similar for both efficacious and non-efficacious ideal Stirling engines. In both cases, as workspace charge is increased, shaft output either increases without bound or increases to a maximum and then decreases to zero.

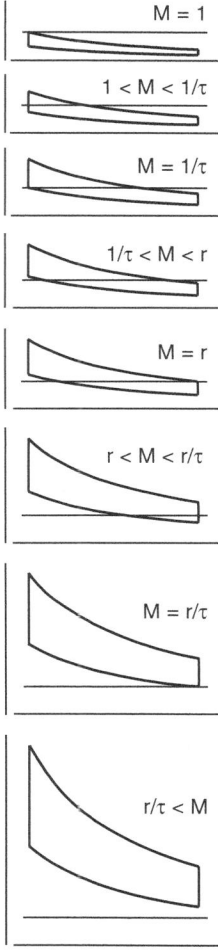

Figure 6.3 Pressure–volume diagrams of a non-efficacious Stirling cycle with fixed buffer pressure shown at various workspace charge levels M. For these cycles, the temperature ratio $\tau = 1/2$, and the compression ratio $r = 5/2$.

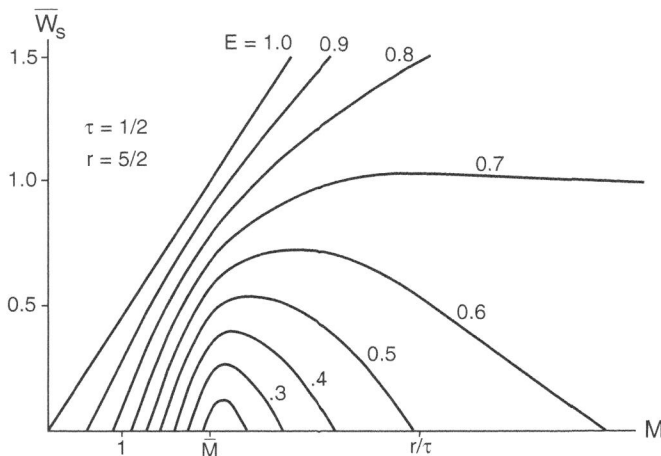

Figure 6.4 Normalized cyclic shaft work $\overline{W}_s(M)$ versus charge mass factor M for a non-efficacious ideal Stirling engine having temperature ratio $\tau = 1/2$ and compression ratio $r = 5/2$. Graphs are shown for constant mechanism effectiveness values E ranging from 1.0 down to 0.2.

It is interesting to note from the graphs in Figures 6.2 and 6.4 that the maximum shaft output occurs at a charge level M greater than $\overline{M}$ and sometimes below r/τ. These are the two easiest charge levels to obtain and maintain in practice. As defined above, $\overline{M}$ is the charge factor for which the average cycle pressure equals the buffer pressure. As pointed out in Chapter 3, this is the charge level that is more or less automatically established and maintained in steady state operation of a Stirling engine not specially equipped with check valves or seals that have a preferred direction of leakage.

The other charge level of practical interest is $M = r/\tau$, the level where the engine workspace cycle is buffered at the minimum cycle pressure. This is also easily obtained in practice by fitting a check valve from the buffer space into the workspace. This was common practice on old hot air water-pumping engines.

Which of these two to choose to obtain the highest mechanical efficiency is obviously $\overline{M}$ because it is known from Chapter 3 that the average cycle pressure p_m of an ideal Stirling turns out to be very close to the optimum buffer pressure p^*. Therefore, the mechanical efficiency at the charge level $\overline{M}$ will be equal to, or very nearly equal to, the maximum possible, namely, r_{ims}.

On the other hand, which of the two workspace charge levels to choose to obtain the maximum shaft output against a fixed buffer pressure is not obvious. The question is given an engine of compression ratio r and mechanism with effectiveness E, running at a temperature ratio τ against a fixed buffer pressure, which workspace charge level, $\overline{M}$ or r/τ, will give the greater shaft output? Or, in very practical terms, shall a check valve into the workspace be fitted or not? The answer can easily be calculated from the equations derived up to this point, with care taken to use the proper expression for the efficacious and non-efficacious cases.

To illustrate, consider ideal Stirlings operating over the three pairs of temperatures given in Table 6.1. The first pair of temperatures is representative of recent high performance Stirling engines (West, 1986). At the opposite extreme, the last pair is the temperature extremes of a small Ringbom engine operating on solar energy (Senft, 1993a). The middle pair was chosen as representative of many model engines and perhaps some of the early hot air engines. In any event, the three corresponding temperature ratios cover most of the range over which real engines operate.

Since most practical Stirling engines have a compression ratio well below 2.5, a typical engine operating at a temperature ratio of 0.33 will

Table 6.1 *Common Stirling engine operating temperatures.*

Cold End Temperature (°C)	Hot End Temperature (°C)	Temperature Ratio τ
50	700	0.33
70	300	0.60
35	90	0.85

be efficacious. The first six lines of Table 6.2 show that the effect of fitting a check valve to raise the charge in such an engine can be quite beneficial. The last column gives the ratio of shaft work at the higher charge level r/τ relative to the shaft work at $\overline{M}$ calculated using the formulas derived for $\overline{W_s}$.

The two columns preceeding the last show the mechanical efficiency at each of the two charge levels calculated using the formulas for $\overline{W_s}$ and $\overline{W}$ in this way:

$$\eta_m\big|_M = \frac{\overline{W_s}(M)}{\overline{W}(M)}.$$

Even though charging to r/τ more than doubles the shaft output, the penalty in mechanical efficiency is not very great in the cases with a

Table 6.2 *Comparison of mechanical efficiency and shaft output of ideal Stirling engines at two workspace charge levels and fixed buffer pressure.*

τ	r	E	$\eta_m\big|_{\overline{M}}$	$\eta_m\big|_{r/\tau}$	$\dfrac{\overline{W_s}(r/\tau)}{\overline{W_s}(\overline{M})}$
0.33	2.0	0.9	0.9	0.87	2.70
0.33	2.0	0.8	0.8	0.76	2.58
0.33	2.0	0.7	0.7	0.60	2.39
0.33	1.5	0.9	0.9	0.88	2.40
0.33	1.5	0.8	0.8	0.76	2.33
0.33	1.5	0.7	0.7	0.64	2.23
0.60	2.0	0.8	0.79	0.61	1.44
0.60	2.0	0.7	0.68	0.39	1.07
0.60	2.0	0.6	0.57	0.15	0.50
0.60	1.5	0.8	0.8	0.68	1.38
0.60	1.5	0.7	0.7	0.51	1.17
0.60	1.5	0.6	0.6	0.32	0.85
0.85	1.5	0.9	0.85	0.69	1.07
0.85	1.5	0.8	0.70	0.35	0.66
0.85	1.5	0.7	0.54	—	—
0.85	1.1	0.9	0.9	0.84	1.07
0.85	1.1	0.8	0.8	0.68	0.97
0.85	1.1	0.7	0.7	0.51	0.83

mechanism effectiveness of 0.8 or 0.9. For example, the fifth line of the table shows that fitting a check valve into the workspace of a hot engine with a compression ratio of 1.5 and a mechanism of 80% effectiveness would multiply shaft output by 233% and reduce the potential mechanical efficiency by only 4 percentage points. This is something good to know.

Also good to know is what the tenth line contains. If the same engine is operated at a more modest temperature ratio, $\tau = 0.6$, the shaft output at the higher workspace charge is only 138% of that at the lower charge level, and the penalty in mechanical efficiency is quite high, being reduced from 80% to 68%. Furthermore, if the mechanism effectiveness were to degrade to, say, $E = 60\%$, the twelfth line shows that output would be less at the higher charge than at the lower and the engine would be laboring at only 32% mechanical efficiency.

The last group of lines is interesting for lower temperature differential engine enthusiasts. The fifteenth line shows an engine that will not run at all at the higher charge level. The last two lines show that a compression ratio reduced from 1.5 to 1.1 gives improved mechanical efficiency but that again there is a negative effect on output from the higher workspace charge.

With the equations given here, any desired specific example can be calculated to determine the effects of workspace pressurization on ideal Stirling engines. Although the numbers given here apply to the ideal case, in view of the Stirling Comparison Theorem of Chapter 3, all engines will follow the trends that these numbers show in a general way.

7

CROSSLEY–STIRLING ENGINES

Crossley cycles are described by two isometric processes and two poly-tropic processes of the same kind. The ideal Stirling cycle and the two-stroke Otto, or so-called *adiabatic Stirling*, are special cases. These two cases in fact bracket the spectrum of the four-step cycles that appear to be reasonable idealizations of the actual cycle of real Stirling engines.

Although the ideal Stirling cycle yields the best case analysis, it is a grand idealization of the actual state of affairs in real engines. The isothermal processes present the chief difficulty because of limited heat transfer rates in a real engine. A more realistic model is one in which the isothermal expansion and compression occur at tempera-tures somewhat displaced from the maximum and minimum engine hardware temperatures; this would model the temperature differen-tial that is necessary to drive the heat transfer to and from the engine gas. This is treated in detail in Chapter 11. In many real engines the expansion and compression processes for the most part occur in engine spaces that have relatively little heat transfer area. Thus, it seems that the expansion and compression processes might be closer to adiabatic than to isothermal. Therefore, using the two-stroke Otto cycle has been advocated as a more faithful, but still idealized, cycle for representing real Stirling engines.

In fact, the actual gas processes may be anywhere between isother-mal and adiabatic. In all engines, some heat transfer will occur between the cylinder walls and the working gas. This effect prevails in smaller or slower running engines but diminishes as size or speed grows. High performance engine designs include effective heat exchangers. Within these components, the engine gas behaves more nearly isothermally. This effect combined with more nearly adiabatic processes in other engine spaces can be altogether approximately modeled as a process between the two extremes. It could also happen in an unusual situation that the compression or expansion processes would go sub-adiabatic, as when, for example, displacer motion is so imperfect that a good share of the expansion would take place in the cold section of the engine. However, the range from isothermal to adiabatic covers the cases of most interest.

CROSSLEY CYCLES

Crossley cycles are composed of two isometric and two polytropic processes of the same index (Kolin, 1972). A *polytropic* process is one throughout which the following pressure–volume relation holds:

$$p\,V^\lambda = K$$

where λ and K are constants. λ is called the *index* of the process. Figure 7.1 shows a representative Crossley cycle for $\lambda > 1$.

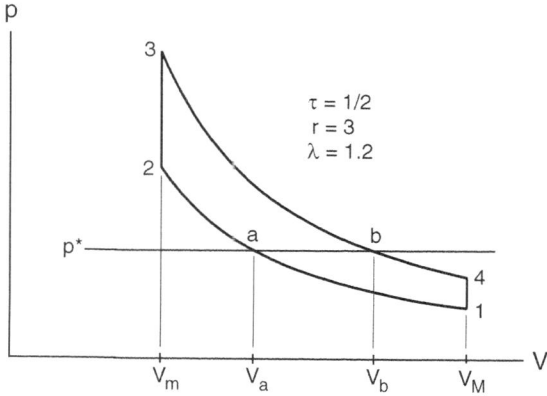

Figure 7.1 The Crossley cycle.

When $\lambda = 1$, the polytropes are isotherms and the cycle is the ideal Stirling. When $\lambda = \gamma = c_p/c_v$, the polytropes are adiabats and the cycle is the two-stroke Otto, which will be henceforth referred to as the *adiabatic Stirling* (following established parlance among Stirling engine workers and to avoid any possible confusion with the four-stroke Otto cycle). Figure 7.1 sets the notation to be used in the following analysis.

CROSSLEY CYCLE ANALYSIS

For an ideal gas, the pressure–volume relation over the compression process 1-2 may be expressed as

$$p\,V^{\lambda} = m\,R\,T_C\,V_M^{\lambda-1} \tag{7.1}$$

and over the expansion 3-4 the relation is

$$p\,V^{\lambda} = m\,R\,T_H\,V_m^{\lambda-1}. \tag{7.2}$$

It follows that the indicated work W of the cycle can be expressed as

$$W = \frac{1}{\lambda - 1}\left(p_1 V_1 - p_2 V_2 + p_3 V_3 - p_4 V_4\right). \tag{7.3}$$

Equations (7.1) and (7.2) yield the following:

$$p_1 V_1 = m\,R\,T_H\,\tau \tag{7.4}$$

$$p_2 V_2 = m\,R\,T_H\,\tau\,r^{\lambda-1} \tag{7.5}$$

$$p_3 V_3 = m\,R\,T_H \tag{7.6}$$

$$p_4 V_4 = m\,R\,T_H\,r^{1-\lambda} \tag{7.7}$$

where $\tau = T_1/T_3 = T_C/T_H$ and $r = V_M/V_m$ are, as before, the ratios of the temperature and volume extremes. Note that $V_1 = V_4 = V_M$ and $V_2 = V_3 = V_m$. Using relations (7.4)–(7.7) in (7.3) gives the following formula for the indicated work of a Crossley cycle:

$$W = \frac{mRT_H}{\lambda - 1}\left(r^{\lambda-1} - 1\right)\left(r^{1-\lambda} - \tau\right) \tag{7.8}$$

which is valid for $\lambda > 1$. Actually, formula (7.8) is also valid for the limiting case $\lambda = 1$ in the sense that on letting $\lambda \to 1^+$ the familiar result (4.2) for the ideal Stirling cycle is obtained, namely,

$$W = m R T_H (1 - \tau) \ln r.$$

In this analysis, the interest is in those Crossley cycles that represent engine rather than heat pump operation. This is the case exactly when $p_2 \leq p_3$ and $p_1 \leq p_4$. Using relations (7.4)–(7.7), both of these inequalities reduce to

$$\tau \leq r^{1-\lambda} \tag{7.9}$$

which will be assumed to hold throughout the following analysis.

FORCED WORK OF THE CROSSLEY CYCLE

A Crossley cycle in which $p_2 \leq p_4$ is efficacious, because a buffer pressure level chosen anywhere between these two pressures will intersect both of the isometric lines of the cycle and the forced work will be zero. Using relations (7.5) and (7.7), the efficacious Crossley cycle is characterized by the inequality

$$\tau \, r^{2\lambda - 1} \leq 1.$$

Figure 7.1 portrays the non-efficacious case, $\tau \, r^{2\lambda - 1} > 1$. Let p^* denote the optimum buffer pressure for this case. Expressions for the forced work occurring on compression and expansion are, respectively,

$$\frac{1}{\lambda - 1}(p_2 V_2 - p^* V_a) - p^*(V_a - V_2) \tag{7.10}$$

and

$$p^*(V_4 - V_b) - \frac{1}{\lambda - 1}(p^* V_b - p_4 V_4) \tag{7.11}$$

That the buffer pressure p^* is optimal means that the total forced work is a minimum for that buffer pressure. As in Chapter 3, this is the case if and only if

$$V_M - V_b = V_a - V_m \tag{7.12}$$

From (7.1) and (7.2) it follows that

$$\left(\frac{V_a}{V_b}\right)^\lambda = \tau r^{\lambda-1} \quad \text{or} \quad V_a = \tau^{1/\lambda} r^{(\lambda-1)/\lambda} V_b$$

which combined with (7.12) gives

$$\left(\tau^{1/\lambda} r^{(\lambda-1)/\lambda} + 1\right) V_b = V_m(r+1)$$

so that

$$V_b = V_m \frac{r+1}{\tau^{1/\lambda} r^{(\lambda-1)/\lambda} + 1}.$$

Raising both sides to the power λ and multiplying by p^* gives

$$p^* V_b^\lambda = p^* V_m^\lambda \left(\frac{r+1}{\tau^{1/\lambda} r^{(\lambda-1)/\lambda} + 1}\right)^\lambda \tag{7.13}$$

Equation (7.2) shows

$$p^* V_b^\lambda = m R T_H V_m^{\lambda-1}.$$

Combining this with (7.13) and solving for p^* yields the following formula for the optimum Crossley cycle buffer pressure:

$$p^* = \frac{m R T_H}{V_m} \left(\frac{\tau^{1/\lambda} r^{(\lambda-1)/\lambda} + 1}{r+1}\right)^\lambda. \tag{7.14}$$

The total forced work of the cycle is the sum of (7.10) and (7.11), which at the optimum buffer pressure simplifies to

$$W_- = \frac{1}{\lambda-1} \left(p_2 V_2 + p_4 V_4 - p^*(V_a + V_b)\right) \tag{7.15}$$

using Equation (7.12). Using (7.12) again, in the form $V_a + V_b = V_M + V_m$ in the last term of (7.15), and also using (7.5) and (7.7) on the two preceding terms, and finally using the expression (7.14) for p^*, Formula (7.15) can be written as

$$W_- = \frac{m R T_H}{\lambda-1} \left[\tau r^{\lambda-1} + r^{1-\lambda} - \frac{\left(\tau^{1/\lambda} r^{(\lambda-1)/\lambda} + 1\right)^\lambda}{(r+1)^{\lambda-1}}\right]. \tag{7.16}$$

Dividing (7.16) by (7.8) and making use of the Fundamental Efficiency Theorem (3.2) gives the following result.

MECHANICAL EFFICIENCY OF CROSSLEY ENGINES

The mechanical efficiency of an optimally buffered Crossley engine with constant mechanism effectiveness E, temperature ratio τ, compression ratio r, and polytropic index λ is

$$\eta_{mc}(E, \tau, r, \lambda) = E - \left(\frac{1}{E} - E\right) S_c(\tau, r, \lambda)$$

where   (7.17)

$S_c(\tau, r, \lambda) = 0$ if $\tau r^{2\lambda-1} \leq 1$, but if $\tau r^{2\lambda-1} > 1$ then

$$S_c(\tau, r, \lambda) = \frac{\left(\tau r^{\lambda-1} + r^{1-\lambda}\right)(r+1)^{\lambda-1} - \left(\tau^{1/\lambda} r^{(\lambda-1)/\lambda} + 1\right)^{\lambda}}{(r+1)^{\lambda-1}\left(r^{\lambda-1} - 1\right)\left(r^{1-\lambda} - \tau\right)}.$$

Although somewhat complex in appearance, the formulas above are exact expressions covering all Crossley engines. Note that Inequality (7.9) ensures that the last factor in the denominator is positive. It is an interesting calculus exercise to verify that the formula for $S_c(\tau, r, \lambda)$ is valid for the Stirling case (3.5)

$$S_c(\tau, r, 1) = \lim_{\lambda \to 1^-} S_c(\tau, r, \lambda) = S(\tau, r).$$

The next three figures show plots of Crossley mechanical efficiency η_{mc} as a function of compression ratio r. The engines of Figure 7.2, with

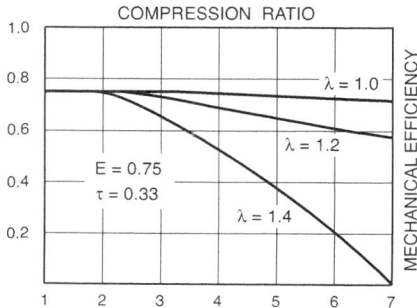

Figure 7.2 Mechanical efficiency versus compression ratio for Crossley engines with E = 0.75 and temperature ratio of 0.33.

$\tau = 0.33$, have a good strong temperature differential between which to work but a rather mediocre mechanism of effectiveness $E = 0.75$. Efficiency plots are shown for three polytropic indices ranging from isothermal $\lambda = 1$ to adiabatic $\lambda = 1.4$ (for diatomic working gas) with $\lambda = 1.2$ chosen in between. For hot engines such as this, there is not much difference in mechanical efficiency between the indices for the range of compression ratios below 2.5.

The situation is notably different for the engines of Figure 7.3, which have less of a temperature difference with which to work, $\tau = 0.60$, even though they have a slightly better mechanism section, $E = 0.80$. At a compression ratio of, say, 2 to 1, the difference in efficiency between an isothermal and an abiabatic cycle is significant.

The engines in Figure 7.4 have a still lower temperature differential, $\tau = 0.85$, and an even better mechanism effectiveness of $E = 0.85$. Here a high penalty is incurred in mechanical losses when departing from the ideal Stirling cycle, and performance is quite sensitively related to compression ratio. For example, at a compression ratio of 1.5, the difference between the isothermal and adiabatic cycle is the difference between running at a decent mechanical efficiency of around 78% and not running at all. Note that in all three cases mechanical efficiency decreases with increasing compression ratio once past the efficacious limit.

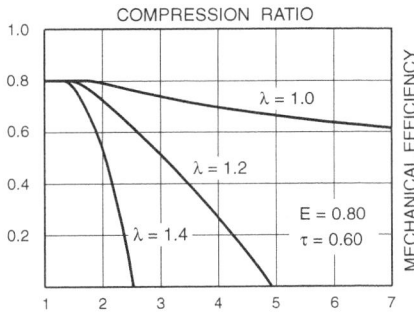

Figure 7.3 Mechanical efficiency versus compression ratio for Crossley engines with $E = 0.80$ and temperature ratio of 0.60.

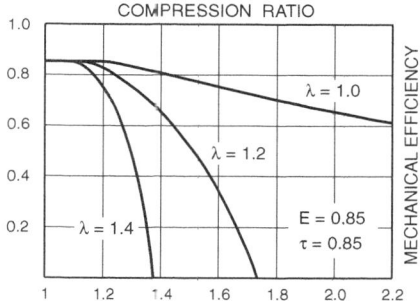

Figure 7.4 Mechanical efficiency versus compression ratio for Crossley engines with E = 0.85 and temperature ratio of 0.85.

THE SWEPT VOLUME RATIO PROBLEM

A basic problem encountered in the early stages of designing a Stirling engine is deciding upon the ratio of piston swept volume to displacer swept volume. While a third-order computer simulation can be brought to bear on the problem, it can do this for only one engine at a time and it is possible only once a good portion of the engine detail has been specified. Having a general intuitive guide would be helpful to start the design process in the first place. This is certainly the case when nontypical or novel engine concepts or applications are under consideration, as, for example, for low-temperature differential operation.

Through clever and intricate mathematical analysis, G. Schmidt (1871) obtained closed-form expressions for the indicated cyclic work of Stirling engines with isothermal spaces and sinusoidal piston and displacer motion. Following the advent of the computer, Schmidt's results have been used by Finkelstein (1960), Kirkley (1962), Walker (1962), and others to numerically determine optimum swept volume ratios based upon indicated cyclic work. For high-temperature differential single-cylinder (beta type) Stirling engines, the results show that a swept volume ratio of around 1 is near optimal. However, these results are based upon indicated rather than upon shaft work output. Results of the previous chapters have shown that increases in indicated work do not always result in increases in shaft output.

The Schmidt analysis has been more recently combined with the Fundamental Efficiency Theorem (3.2) to numerically determine the swept volume ratios and phase angles that maximize cyclic shaft work (Senft, 2001, 2002). This work has revealed that optimum engine geometries are not simply thermodynamically determined, but are heavily dependent upon the underlying effectiveness of the engine mechanism. Most significantly, it has been found that maximum shaft output occurs at smaller swept volume ratios than does maximum indicated work. Indeed, it was found that some indicated work optima yield Schmidt engines that cannot produce any positive shaft output, that is, they cannot even run themselves. Chapter 10 covers this new Schmidt optimization in detail.

All Schmidt-based analyses assume isothermal conditions in the engine spaces. The Schmidt scheme loses its mathematical tractability when attempts are made to convert it to one with adiabatic expansion and compression spaces. Computer simulations can be done, but insight is lost in the computational complexity. In order to obtain some intuition of how optimal geometry is affected by non-isothermal conditions, one is forced back to the four-step cycle analysis initiated above.

The most meaningful and appropriate way of formulating the swept volume ratio problem for beta-type Stirling engines is the following:

SWEPT VOLUME RATIO PROBLEM

Given temperature extremes T_C and T_H, mean pressure p_m, constant mechanism effectiveness E, and total engine volume V_M, find the piston swept volume $\Delta V = V_M - V_m$ that yields the maximum cyclic shaft work W_s when the engine is buffered at the optimum pressure p^*.

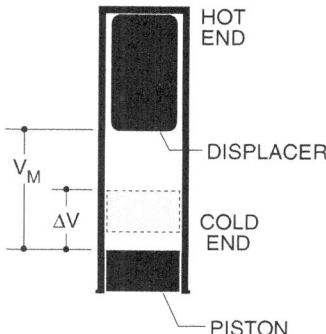

Figure 7.5 Schematic of the swept volume ratio problem.

Figure 7.5 represents the elements of the problem. Only the thermodynamic space need be considered, since the mechanism is characterized by the given effectiveness. The piston is shown in its outermost position, enclosing the maximum workspace volume V_M. Some of the space within the engine capsule is occupied by the displacer, which is shown in its innermost postion; its movement between the hot end and the piston effects the constant volume processes of the Crossley cycle. The clearance space around the displacer is part of the so-called dead space of the engine. In the present idealized analysis it is taken to be negligible. Stirling engines with extended internal heat exchanger surfaces and regenerators can have appreciable dead volume, which will affect the applicability of the results obtained here. However, the main intent here is to obtain some understanding of how non-isothermality can influence optimum engine geometry.

The problem is to determine the piston stroke volume, or equivalently the volume compression ratio, that will result in the maximum shaft work per cycle. If a very small stroke were to be chosen, the engine cycle would be efficacious. Then mechanical efficiency would be as high as possible, but indicated work would be low. At the other extreme, with a very large piston swept volume, indicated work would be high, but mechanical efficiency would be low because of the presence of excessive forced work. Somewhere between these extremes, the product of mechanical efficiency and indicated work would be greatest; this is the point of optimum swept volume or compression ratio that maximizes shaft work output.

CROSSLEY–STIRLING OPTIMUM COMPRESSION RATIOS

According to the statement of the swept volume ratio problem given in the previous section, the object is to find the compression ratio at which the *total-volume mean-pressure specific cyclic shaft work*

$$\hat{W}_s = \frac{W_s}{p_m\,V_M} = \frac{W\,\eta_{mc}}{p_m\,V_M} \tag{7.18}$$

is a maximum.

The mean cycle pressure is calculated using Formulas (1.1) and (1.2) in Defintions (3.8) with the compression and expansion pressure functions for Crossley cycles given by (7.1) and (7.2); the result is

$$p_m = \frac{r}{2\,V_M\,(r-1)}\,\frac{m R T_H}{\lambda-1}\left(r^{\lambda-1}-1\right)\left(r^{1-\lambda}+\tau\right).$$

Substituting this into the right-hand member of (7.18) and making use of (7.8) yields

$$\hat{W}_s = \frac{2(r-1)\left(r^{1-\lambda}-\tau\right)}{r\left(r^{1-\lambda}+\tau\right)}\,\eta_{mc} \tag{7.19}$$

where $\eta_{mc} = \eta_{mc}(\mathrm{E}, \tau, r, \lambda)$ is given by (7.17). Expression (7.19) is valid for $\lambda > 1$. The form of (7.19) for the ideal Stirling case[†] when $\lambda = 1$ was obtained directly in Chapter 4, namely,

$$\hat{W}_s^* = 2\frac{(r-1)(1-\tau)}{r(1+\tau)}\,\eta_{ms}. \tag{4.4}$$

Expressions (7.19) and (4.4) show that the optimum compression ratio r values will depend only upon the temperature ratio τ in addition to the polytropic index λ and the mechanism effectiveness E.

Figures 7.6, 7.7, and 7.8 show plots of specific shaft work given by Equations (7.19) and (4.4) versus r for $\lambda = 1.0, 1.2,$ and 1.4.

[†] It is noteworthy that Expression (4.4) is actually the limiting case of (7.19) since $\eta_{mc}(\mathrm{E}, \tau, r, 1) = \lim_{\lambda \to 1^+} \eta_{mc}(\mathrm{E}, \tau, r, \lambda) = \eta_{ms}(\mathrm{E}, \tau, r)$, as indicated earlier.

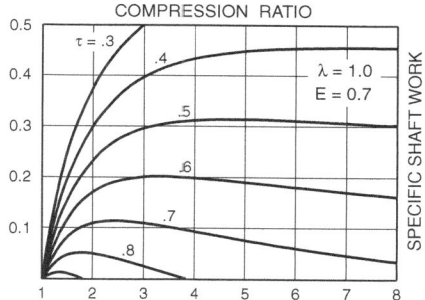

Figure 7.6 Graphs of specific shaft work versus compression ratio for ideal Stirling engines.

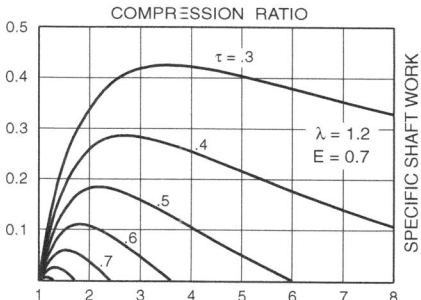

Figure 7.7 Specific shaft work for Crossley engines with $\lambda = 1.2$.

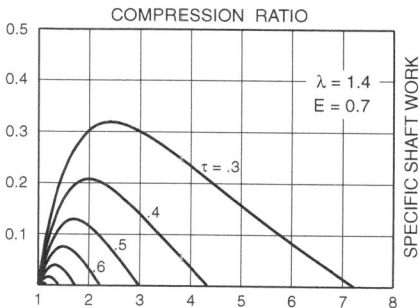

Figure 7.8 Specific shaft work graphs for adiabatic Stirling engines.

Figures 7.6–7.8 Graphs of specific shaft work versus compression ratio of three types of Crossley engines for temperature ratios τ ranging from 0.3 to 0.9 in increments of 0.1. Figure 7.6 is for the the polytropic index $\lambda = 1.0$ (ideal Stirling). Figure 7.7 is for $\lambda = 1.2$, and Figure 7.8 is for $\lambda = 1.4$ (Otto or adiabatic Stirling). The mechanism effectiveness is $E = 0.7$ in all these graphs.

In each illustration, τ ranges from 0.3 to 0.9 in increments of 0.1. The mechanism effectiveness is the same in all, namely E = 0.7. Each figure employs the same scale so that direct comparisons can be made.

One observes in these graphs a dramatic decrease in engine output as the polytropic index λ increases. Taking as an example, $\tau = 0.5$, the peak specific cyclic shaft work in the isothermal case is about 0.32, as shown in Figure 7.6. In the intermediate case with $\lambda = 1.2$, the peak shaft work for the same temperature ratio drops almost in half to around 0.18. For the adiabatic case, Figure 7.8, the output peak is at about 0.13 for the same $\tau = 0.5$ engine.

In the adiabatic Stirling, the optimum compression ratio is about 2.4 when $\tau = 0.3$. As already mentioned, this temperature ratio is about the best that can be attained in practice with reasonable choices of hot end materials and typical sink temperatures. At $\tau = 0.4$ in the adiabatic Stirling, the best compression ratio is about 2.0. The graphs show how optimum compression ratio decreases with further increases in τ.

The situation for optimum compression ratios is similar in the cases with lower polytropic index, but the optimums are larger. In the case of the ideal Stirlings in Figure 7.6, the theoretical optimums are quite high for the hotter engines and well beyond what is attempted in practice. It is also of interest that in the lower λ engines near peak output is not as sensitive to compression ratio as it is in the adiabatic case.

It is interesting to compare the shaft work plots with indicated work. The appropriate indicated work measure for this purpose would be the *mean-pressure total-volume specific indicated cyclic work*

$$\hat{W} = \frac{W}{p_m V_M} \tag{7.20}$$

where W and p_m are as before. Figures 7.9, 7.10, and 7.11 show $\hat{W}$ as a function of compression ratio r for the same cases shown in Figures 7.6, 7.7, and 7.8, respectively.

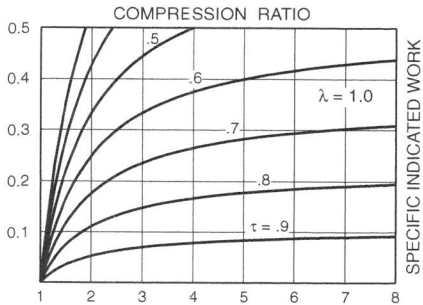

Figure 7.9 Graphs of specific indicated work versus compression ratio for ideal Stirling engines.

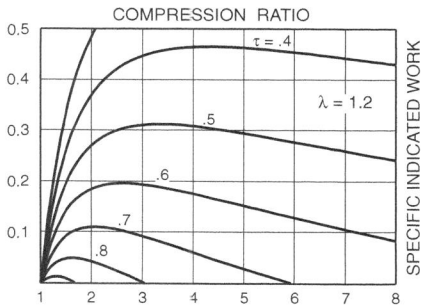

Figure 7.10 Specific indicated work for Crossley engines with $\lambda = 1.2$.

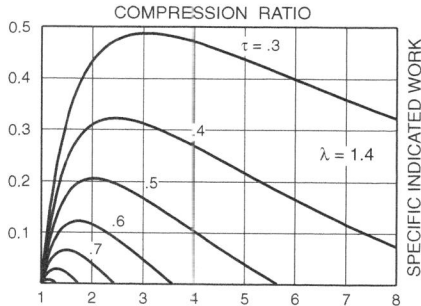

Figure 7.11 Specific indicated work graphs for adiabatic Stirling engines.

Figures 7.9–7.11 Graphs of specific indicated work versus compression ratio for the three types of Crossley-Stirling engines whose shaft work output is depicted in Figures 7.6–7.8. The scales are the same in all of the figures.

Looking first at the adiabatic case, comparing Figures 7.8 and 7.11, one sees that the losses suffered in the mechanical section of the engine not only decrease output of course, but also decrease compression ratio optima. This effect of mechanical losses lowering the optimum compression ratio is even greater in the mixed case, $\lambda = 1.2$, as comparison of Figures 7.7 and 7.10 shows.

For the ideal, i.e., the isothermal Stirling, there are no optimum compression ratios based on indicated work, as Figure 7.9 illustrates, but they do appear in Figure 7.6, where mechanical losses are taken into account. This explains why the optimum compression ratios for the shaft work of ideal Stirlings are so large, as already noted.

CONCLUSIONS

As detailed in the opening paragraphs of this chapter, the goal of the analysis here was to acquire some insight into how non-isothermal processes affect engine performance. In summary, the following results stand out:

- A high penalty in mechanical efficiency is exacted for departing from isothermal and moving toward adiabatic processes. There is also a high penalty in specific indicated work. The two effects combine to significantly reduce shaft output.

- All Crossley cycles show maxima of specific indicated work with respect to compression ratio when $\lambda > 1$. The specific indicated work of the ideal Stirling, $\lambda = 1$, monotonically increases to a limiting value of $2(1 - \tau)/(1 + \tau)$ as $r \to \infty$.

- All Crossley cycles, including the ideal isothermal Stirling, have maxima for specific shaft work with respect to compression ratio. The effect of a non-perfect engine mechanism (i.e., $E < 1$) is to decrease the optimum compression ratio.

- For λ equal to 1, the optimum compression ratio for shaft work occurs beyond the efficacious range. In other words, optimal ideal Stirling engines are non-efficacious.

- The optimum compression ratio for adiabatic Stirling engines based on brake output is below 2.5 for practical temperature ratios. The optimum compression ratio decreases as engine operating temperature differential decreases and occurs in the efficacious range. This shows that the practical Stirling is intrinsically at its best with low compression ratios. Too high a compression ratio in a low ΔT engine could preclude operation altogether. On the other hand, because the maximum output is attained at low compression ratio values in the adiabatic engine, performance is more sensitive to compression ratio in the range below the optimum value.

- The intrinsically low optimum compression ratios of adiabatic Stirlings found here agree well with those based upon thermal efficiency considerations in research carried out by Rallis and Urieli (1976). For imperfect regeneration in an engine with temperature ratio of $\tau = 0.35$, they found that maximum thermal efficiency occurs at a compression ratio of around 2. This is in good general agreement with the results of this chapter based upon shaft output.

8

GENERALIZED ENGINE CYCLES
AND VARIABLE BUFFER PRESSURE

Previous chapters covered engines with p-V cycles that could be described by two continuous pressure functions $p_e(V) \geq p_c(V)$ over the interval $[V_m, V_M]$ of volume variation of the engine. Such cycles were termed *regular*. Regularity is not an overly restrictive limitation, but some engine cycles are not that simple. Figure 8.1 shows examples of non-regular cycles having intermediate constant volume processes, pumping loops, and multiple strokes or volume reversals.

PARAMETRIC REPRESENTATION

A general engine cycle can be mathematically described as a continuous piecewise smooth closed curve in the V-p plane, that is, a closed curve that is the union of a finite number of endpoint-connected smooth curves (Buck, 1965). Such curves can be described parametrically as

$$\Upsilon = (V, p) = (\alpha(s), \beta(s)), \quad s \in J = [a, b] \tag{8.1}$$

where J is a closed bounded interval of length equal to the least common period of the functions α and β. In this setting of course, V and p are positive at all times. The *indicated cyclic work* of the engine is by definition

$$W = \int_{\Upsilon} p \, dV = \int_a^b \beta(s)\alpha'(s) \, ds. \tag{8.2}$$

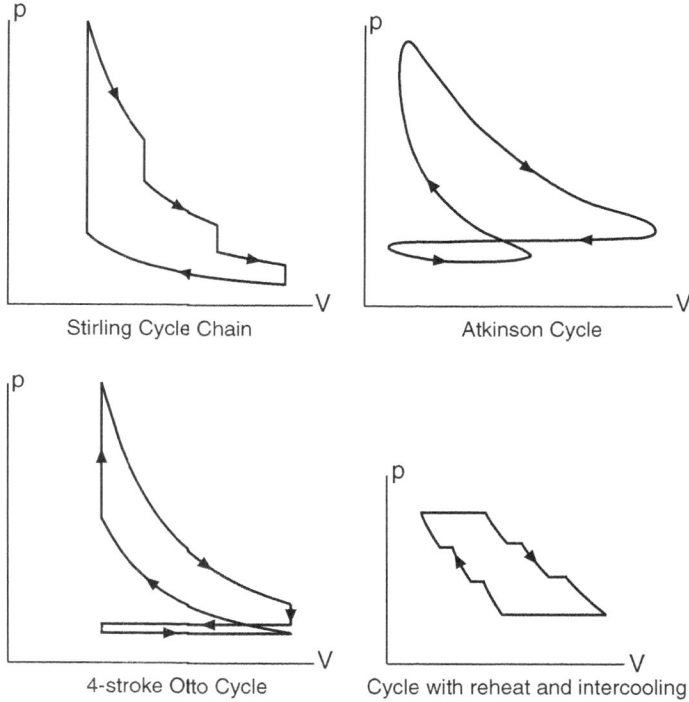

Figure 8.1 Examples of general engine cycles.

For regular engine cycles, the indicated work is simply the area inside the cycle. For general cycles, areas enclosed by the cycle that are bounded by a counterclockwise-oriented subcycle are to be subtracted from the subcycle areas with a clockwise boundary. Thus, in the Atkinson cycle shown in Figure 8.1, the area within the small pumping subcycle is to be subtracted from the area in the large upper subcycle. The same is true for the idealized four-stroke Otto cycle (Taylor, 1966).

Note that in cases where the working fluid is not homogeneous, if p is the pressure at the face of the piston, then W still represents the cyclic work done *on the piston*. In addition, it is not necessary that V be the actual volume of the workspace in the calculation. It is only necessary that the total volume variation of the fluid effected by the piston be represented parametrically. However, V will be referred to as the workspace volume and p as the workspace pressure because that is usually the case in fact or is a reasonable approximation. Similar considerations apply to the other work quantities defined below.

The *absolute expansion work* and *absolute compression work* of cycles in the present general setting can be expressed as

$$W_e = \int_\Upsilon p\,(dV)^+ = \int_a^b \beta(s)\big(\alpha'(s)\big)^+ ds$$

$$W_c = \int_\Upsilon p\,(dV)^- = \int_a^b \beta(s)\big(\alpha'(s)\big)^- ds. \tag{8.3}$$

The plus and minus superscripts in (8.3) denote the *positive part* and *negative part* functions as described in Chapter 5. The identities (8.3) and (5.2) show indeed that the difference between the absolute expansion and compression works is the indicated work:

$$W_e - W_c = \int_\Upsilon p\,dV^+ - \int_\Upsilon p\,dV^- = \int_\Upsilon p(dV^+ - dV^-)$$

$$= \int_\Upsilon p\,dV = W.$$

Parentheses were omitted from around the dV factors here for brevity, and will be henceforth, with the understanding that $dV^+ = (dV)^+$ and $dV^- = (dV)^-$. The absolute expansion and absolute compression works of a nonregular cycle are shown in Figure 8.2.

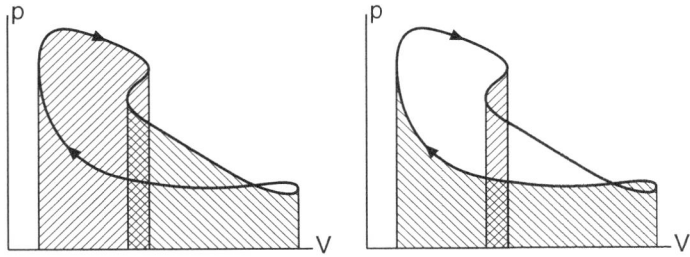

Figure 8.2 A general engine cycle with its absolute expansion and absolute compression works identified by the hatched areas. The double-hatched areas are counted twice.

AVERAGE CYCLE PRESSURES

Characteristic average cycle pressures were defined in Chapter 3 for regular cycles. These concepts can be extended to suit the wider range of cycles now under consideration. First, the *average expansion pressure* is defined as

$$p_{ae} = \frac{1}{\Delta V^+} \int_\Upsilon p \, dV^+ = \frac{W_e}{\Delta V^+}$$

where the denominator is the total volume swept by the piston during all of the expansion subprocesses of the cycle, that is,

$$\Delta V^+ = \int_\Upsilon dV^+.$$

The *average compression pressure* is defined as

$$p_{ac} = \frac{W_c}{\Delta V^-}.$$

Note that $\Delta V^- = \int_\Upsilon dV^- = \int_\Upsilon dV^+ = \Delta V^+$ because Υ is a cycle, i.e., a closed curve. The *average* or *mean cycle pressure* is then defined as

$$p_m = \frac{p_{ae} + p_{ac}}{2}.$$

It is interesting to note that

$$p_m = \frac{W_e + W_c}{|\Delta V|} \tag{8.4}$$

where $|\Delta V| = \int_\Upsilon |dV| = \Delta V^+ - \Delta V^-$ is the total bidirectional piston swept volume. These definitions are direct generalizations of those made earlier for regular cycles (3.8). When the cycle is regular, $\Delta V^+ = V_M - V_m$, p_{ae} is the mean upper pressure p_{mu}, and p_{ac} is the mean lower pressure p_{ml} of the cycle.

VARIABLE BUFFER PRESSURE

Thus far, buffer pressure has been assumed constant. This is exactly the case for typical Otto and Diesel engines and for Stirling engines having a crankcase open to the atmosphere. Furthermore, it is a reasonable simplification for many other engines. A two-stroke internal combustion engine is a good example of this; because of crankcase pumping, the buffer pressure does vary, but the variation is quite small compared to the workspace pressure variation and so can usually be neglected in analyses of mechanical efficiency.

As will be shown in this section, constant buffer pressure represents the best possible case. Sometimes, however, buffer pressure variation is unavoidably substantial and cannot be ignored as for example in high mean pressure Stirling engines where buffer volume comes at a high structural cost. A hypothetical example of buffer pressure variation is shown in Figure 8.3. This kind of variation, which is typical, adversely affects mechanical efficiency as will be shown later. Moreover, in practice, confined buffer gas can experience a non-negligible cyclic energy loss due to transient heat transfer to the container walls (West, 1986). The buffer pressure trace when this type of loss occurs is illustrated in Figure 8.4. Here the buffer pressure is plotted against the workspace volume, that is, the illustration depicts both the workspace pressure and the corresponding bufferspace pressure as the engine working volume varies. The orientation shown for each of the curves maintains this correspondence on both the compression and expansion processes of the engine space.

Mathematically this is automatically taken care of if the buffer pressure is represented by a function of the same parameter as used for the workspace pressure and volume. Thus, a complete engine and buffer combination can be specified by a three-component set of functions:

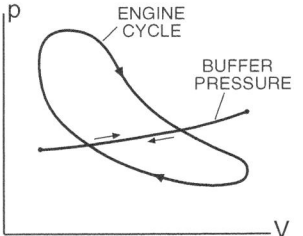

Figure 8.3 An engine cycle with variable buffer pressure.

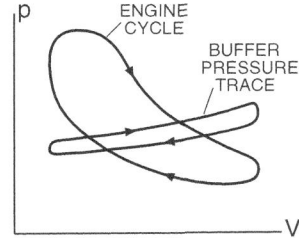

Figure 8.4 An engine cycle with a buffer that dissipates energy.

$$\Upsilon = (V, p, p_b) = (\alpha(s), \beta(s), \gamma(s)).$$

The physical interpretation is that

$$V = \alpha(s), \quad p = \beta(s), \quad s \in J$$

is the volume–pressure cycle of the workspace, while $p_b = \gamma(s)$ is the pressure of the buffer gas at the workspace volume $V = \alpha(s)$.

Note then that the curve $(V, p_b) = (\alpha(s), \gamma(s)), s \in J$ is not actually the volume–pressure cycle of the buffer gas, but a *trace* of the buffer pressure plotted against the workspace volume. It is closely related to the cycle of the buffer gas, however, in that the working piston is usually the sole volume-varying element of the bufferspace, and hence the volume variation of the bufferspace is exactly opposite that of the workspace. Therefore the area enclosed by the buffer pressure trace represents the work *done on* the buffer fluid by the piston, with the usual convention that an area enclosed by a clockwise-oriented boundary is taken as positive, and when bounded by a counterclockwise-oriented curve is taken as negative. Thus, the example of Figure 8.4 shows a positive amount of work done on the buffer fluid and dissipated through thermal interaction with the bufferspace walls. In general, the *cyclic work done on the buffer fluid* is given by

$$W_b = \int_\Upsilon p_b \, dV = \int_a^b \gamma(s)\alpha'(s)ds. \tag{8.5}$$

Note that this definition also includes cases where the buffer has a non-simple trace. In this chapter, it will be assumed that $W_b \geq 0$, that is, that the buffer space does not itself act as a heat engine workspace. The topic of multi-workspace engines will be treated in the next chapter.

BUFFER PRESSURE AND ENERGY TRANSFERS

Figure 8.5 shows an engine cycle and buffer pressure trace with portions labeled according to the direction of positive work transfer from the piston. The parts labeled with a plus sign show where the piston delivers work to the mechanism. These are the portions of the cycle where the workspace pressure is either above the buffer pressure and workspace volume is expanding, or where the workspace is being compressed but its pressure is below the buffer pressure. These are the *efficacious processes* of the cycle, and the piston work delivered during them is called the *efficacious work* of the cycle and denoted by W_+. Conceptually and mathematically the efficacious work here is the same as before for constant buffer pressure, only now the buffer pressure may change with volume:

$$W_+\langle \Upsilon \rangle = \int_\Upsilon [(p - p_b)\, dV]^+ . \tag{8.6}$$

The negative signs in Figure 8.5 show where the mechanism must do work on the piston to force it to carry out the process. The *forced processes* occur whenever a compression is being carried out with the buffer pressure below the workspace pressure, or when an expansion is made to take place with the workspace pressure below the buffer pressure. As before, the *forced work* W_- of the cycle is the positive work that the mechanism must deliver to the piston to carry out the forced processes:

$$W_-\langle \Upsilon \rangle = \int_\Upsilon [(p - p_b)\, dV]^- . \tag{8.7}$$

Figure 8.6 shows the forced work of an engine with dissipative buffer pressure.

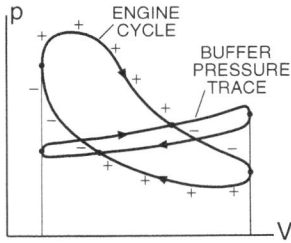

Figure 8.5 The efficacious and forced processes of Figure 8.4 are here identified by the plus and minus signs, respectively.

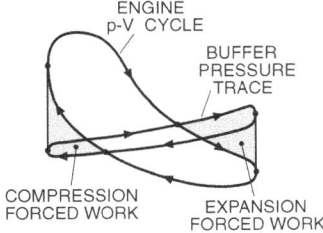

Figure 8.6 The forced work of Figure 8.5 is shown here as the shaded areas.

A fundamental relation between the work quantities described here can be easily deduced using (8.2), (8.5)–(8.7), and the simple identity $z = z^+ - z^-$:

$$
\begin{aligned}
W - W_b &= \int_\Upsilon p\, dV - \int_\Upsilon p_b\, dV = \int_\Upsilon (p - p_b)dV \\
&= \int_\Upsilon [(p - p_b)\, dV]^+ - \int_\Upsilon [(p - p_b)\, dV]^- \\
&= W_+ - W_-
\end{aligned}
\tag{8.8}
$$

MECHANICAL EFFICIENCY

If the effectiveness of an engine mechanism is bounded above by the constant E over the complete cycle of operation, then its shaft work output over a cycle is bounded above by $E\,W_+ - W_-/E$. This follows from the same reasoning that led to Inequality (2.2).[†] Making use now of identity (8.8),

$$
\begin{aligned}
W_s \le EW_+ - \frac{W_-}{E} &= E(W - W_b + W_-) - \frac{W_-}{E} \\
&= E(W - W_b) - \left(\frac{1}{E} - E\right) W_-
\end{aligned}
$$

Dividing by the indicated work W gives the following extension of the Fundamental Efficiency Theorem (3.1).

FUNDAMENTAL EFFICIENCY THEOREM

If the effectiveness ε of an engine mechanism is bounded above by the constant $E \le 1$, that is, if $\varepsilon \le E \le 1$ throughout the cycle, then the mechanical efficiency η_m of the engine satisfies

$$
\eta_m \le E - \left(\frac{1}{E} - E\right)\frac{W_-}{W} - E\frac{W_b}{W}
\tag{8.9}
$$

where W is the indicated work of the cycle, W_- is the forced work, and $W_b \ge 0$ is the buffer gas dissipation. Equality holds if the mechanism effectiveness equals the constant E over the entire cycle.

[†] It also follows from Inequality (A 19) of Appendix A.

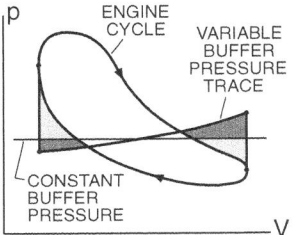

Figure 8.7 The forced work of an engine with constant buffer pressure is shown here as the lighter shaded area. The darker shaded area is additional forced work arising from typical buffer pressure variation.

Inequality (8.9) is identical to the earlier result (3.1) except for an expected additional penalty term corresponding to buffer gas dissipation. When there is no buffer gas loss, that is, when $W_b = 0$, Inequality (8.9) becomes identical to (3.1). The present result, however, applies when buffer pressure is nonconstant.

As already claimed, constant buffer pressure is the most desirable. Variable buffer pressure always entails more mechanical loss, even when there is no dissipation, because of increased forced work. Consider an engine where the bufferspace volume is small enough so that its pressure variation is non-negligible. Its pressure will rise as the workspace expands and will fall when the workspace is compressed. This is illustrated in Figure 8.7. Also shown in the figure for comparison is a centrally placed constant buffer pressure. The forced work for the variable case is greater because the buffer pressure rises as the workspace pressure falls, and vice versa. Looking back at Figure 8.6 now will reveal that if dissipation were also present, the forced work would be greater still. This is why a bufferspace as large as possible, that is, a buffer pressure as constant as possible, yields better mechanical efficiency.

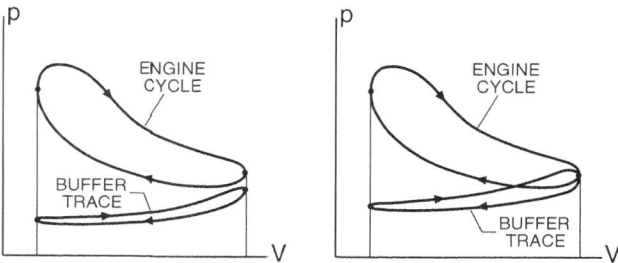

Figure 8.8 Two examples of an engine cycle buffered from below with nonconstant pressure.

PRESSURIZATION EFFECTS

The results of Chapter 5 on pressurization can be extended to cover the more general engine cycles and variable buffer pressure treated in this chapter. The concept of monomorphic cycles applies to buffer pressure traces, and the System Charging Theorem remains valid in this extended context. The theorems on engines buffered from below have also been extended to apply to engines with variable buffer pressure, as illustrated in Figure 8.8. In particular, the Workspace Charging Theorem has also been formulated to apply when buffer pressure is variable. The details of these extended results can be found in the references pertaining to pressurization (Senft, 1988, 1991b).

9

MULTI-WORKSPACE ENGINES
AND HEAT PUMPS

This chapter shows how the previous results can be extended to apply to engines having more than one workspace and also to heat pumps.

MULTI-CYLINDER ENGINES

Most multi-cylinder engines consist of a number of identical single-piston units connected to a common shaft. This is the case, for example, with in-line and "v" arrangements in internal combustion engines and wobble-plate Stirling engines. The individual units usually have identical mechanisms, share a common constant buffer pressure, and have the same pressure–volume cycles on the piston working faces. The units are usually connected to the common shaft with offset phasing to smooth out the working of the composite machine. For this kind of uniform *parallel connection*, the mechanical efficiency is obviously the same as that of the individual units.

In a more general case, the cycles and mechanisms may be different. Consider a parallel connection as in Figure 9.1 with n units, each having a cycle $\Upsilon_i = (V_i, p_i)$ and a mechanism effectiveness at most equal to the constant E_i between 0 and 1 over its respective cycle. Inequality (2.2) applies to each unit, that is,

$$W_s \langle \Upsilon_i \rangle \leq E_i W_+ \langle \Upsilon_i \rangle - (1/E_i) W_- \langle \Upsilon_i \rangle \qquad (9.1)$$

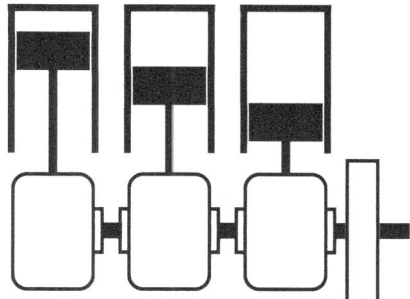

Figure 9.1 A parallel connection of three single-workspace engines.

where the efficacious and forced works W_+ and W_- are defined as before by (8.6) and (8.7). Taking the sum of inequalities (9.1) yields

$$\sum_{i=1}^{n} W_s\langle\Upsilon_i\rangle \leq \sum_{i=1}^{n} E_i\, W_+\langle\Upsilon_i\rangle - \sum_{i=1}^{n} \frac{1}{E_i} W_-\langle\Upsilon_i\rangle.$$

Let $E = \max\{E_i\}$. Then the total net shaft output is bounded above as follows:

$$\sum W_s\langle\Upsilon_i\rangle \leq E\sum W_+\langle\Upsilon_i\rangle - (1/E)\sum W_-\langle\Upsilon_i\rangle. \qquad (9.2)$$

The case of constant buffer pressure is of most applicability for such engines. If buffer pressure is constant, $W = W_+ - W_-$ can be applied to each workspace cycle in the W_+ terms of the right-hand side of (9.2) to yield

$$\sum W_s\langle\Upsilon_i\rangle \leq E\sum W\langle\Upsilon_i\rangle - (1/E - E)\sum W_-\langle\Upsilon_i\rangle.$$

Dividing by the total indicated work $\sum W\langle\Upsilon_i\rangle$ of the cylinder cycles gives an upper bound for the mechanical efficiency of the composite engine:

$$\eta_m\langle\Sigma\Upsilon_i\rangle \leq E - \left(\frac{1}{E} - E\right) \frac{\sum W_-\langle\Upsilon_i\rangle}{\sum W\langle\Upsilon_i\rangle} \qquad (9.3)$$

It is interesting to note that this has the same overall form as in the Fundamental Efficiency Theorem (3.1) for single-workspace engines.

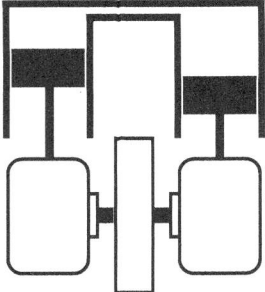

Figure 9.2 An engine with interconnected workspaces.

SPLIT-WORKSPACE ENGINES

The above analysis can also be applied when engine workspaces are interconnected as depicted generically in Figure 9.2. This arrangement was used on the Stirling engine variation invented by A. K. Rider. Another example of the two-piston or so-called alpha type of Stirling engine is shown in Figure 9.3 (Hargreaves, 1991). The cylinder on the right is heated and the left is cooled. Between the two cylinders is a porous matrix, which serves as a regenerator. The pistons move roughly sinusoidally and carry the body of captive gas between the two pistons through an approximate Stirling cycle.

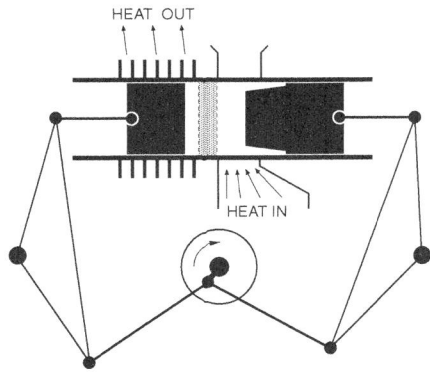

Figure 9.3 The famed W-linkage Stirling engine by Philips.

Idealized p–V cycles Υ_H for the hot piston and Υ_C for the cold piston of an alpha-type Stirling engine can be calculated from the well-known Schmidt model (Kirkley, 1962). The basic assumptions of this model are exact sinusoidal motion of the pistons, a constant phase separation, isothermal hot and cold spaces, an ideal gas working fluid, no leakage, and uniform instantaneous pressure throughout all engine spaces. Equations for the hot and cold volumes and the cycle pressure as functions of crank angle s are as follows:

$$\text{volume of hot space:} \quad V_{hot} = \frac{V_H}{2}(1 - \cos s) + V_{DH}$$

$$\text{volume of cold space:} \quad V_{cold} = \frac{V_C}{2}(1 + \cos(s + \alpha)) + V_{DC}$$

$$\text{cycle pressure:} \quad p = \frac{mR}{V_{hot}/T_H + V_{cold}/T_C}$$

where V_H is the swept volume of the hot piston, V_{DH} the unswept or 'dead' volume in the hot space, V_C the swept volume of the cold piston, V_{DC} the unswept volume in the cold space, α the crank phase angle between the cold and hot piston, m the total mass of gas in the engine, R the gas constant, T_H the temperature of the gas in the hot space, and T_C the temperature of the gas in the cold space.

Examples of hot and cold cylinder cycles Υ_H and Υ_C are shown in Figure 9.4. These were calculated from the above equations for a temperature ratio $T_C/T_H = \frac{1}{3}$, equal piston swept volumes, $V_{DH}/V_H = 0.1 = V_{DC}/V_C$, and a phase angle $\alpha = \frac{\pi}{2}$ between the pistons.

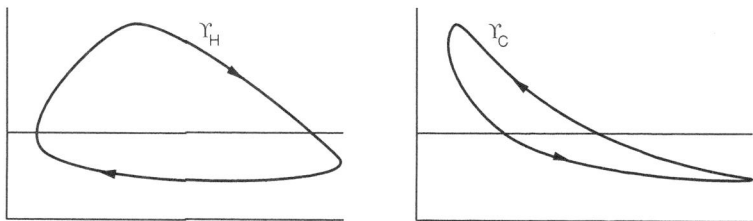

Figure 9.4 Pressure–volume diagrams of the hot and cold cylinders of an alpha-type Stirling engine.

Taking buffer pressure as the geometric or root mean of the maximum and minimum cycle pressures, namely, $p_b = \sqrt{p_{min}p_{max}}$, numerical integration shows that

$$\frac{\sum W_-\langle \Upsilon_i \rangle}{\sum W\langle \Upsilon_i \rangle} = \frac{W_-\langle \Upsilon_H \rangle + W_-\langle \Upsilon_C \rangle}{W\langle \Upsilon_H \rangle + W\langle \Upsilon_C \rangle} = \frac{1.007}{1.241} = 0.811.$$

With a mechanism effectiveness of 90% or less, Inequality (9.3) implies that the mechanical efficiency of this example engine cannot exceed 73%:

$$\eta_m \leq 0.90 - \left(\frac{1}{0.90} - 0.90 \right)(0.81) = 0.729.$$

This is a significant reduction from the 90% mechanism performance level.

It is also a marked reduction from the 84% efficiency level for the beta-type Stirling engine example of Figure 3.1, yet the pressure versus total volume cycles of the two engines are exactly the same. The two engines carry out the same thermodynamic cycle, have the same buffer pressure, and have equally effective mechanisms, but their different mechanical arrangement has a significant effect upon their overall efficiency. Although the mechanical losses associated with the displacer drive of the beta engine have not been taken into account in this comparison, they are relatively small in a well-designed and well-constructed engine.

ENGINES WITH DOUBLE-ACTING PISTONS

Figure 9.5 conceptually represents an engine in which the piston is acted upon from both sides by a variable pressure cycle. There is the *primary cycle* $\Upsilon = (V, p)$ on one side of the piston as usual, and a *secondary cycle* $\Upsilon_b = (V_b, p_b)$ acting on the other side. The secondary cycle may simply be the buffer; both constant and variable buffer pressure cases have been thoroughly treated in the preceeding chapters. Of special interest here is the case when the secondary cycle acts as an engine cycle.

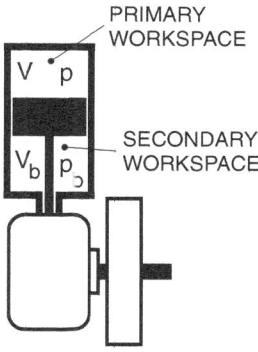

PRIMARY
WORKSPACE

V p

SECONDARY
WORKSPACE

V_b p_b

Figure 9.5 Schematic diagram of a double-acting engine.

If the secondary pressure p_b is indexed with the same parameter as used for the primary cycle as in the previous chapter, then the formulas of that chapter validly apply. A little care needs to be exercised to get the signs associated with the secondary cycle correct; if one adheres to a strict formalism, this is automatically taken care of. This is illustrated in the following verification that the net work done on the piston by the dual force process is indeed equal to the sum of the indicated work of the primary and secondary cycles. Assuming the active areas of both sides of the piston are essentially equal to A, the force acting on the piston is $f = (p - p_b)A$. If x represents the linear position of the piston, then, $A dx = dV$, and so

$$W = \int_J f dx = \int_\Upsilon (p - p_b) dV = \int_\Upsilon p dV - \int_\Upsilon p_b dV$$
$$= \int_\Upsilon p dV + \int_{\Upsilon_b} p_b dV_b = W\langle\Upsilon\rangle + W\langle\Upsilon_b\rangle.$$

The relation $dV = -dV_b$ used in the next to last step of the calculation is a consequence of the fact that $V + V_b$ is ideally constant in double-acting engines. The definitions of the efficacious and forced works of the engine remain exactly as given in equations (8.6) and (8.7).

Indexing the secondary cycle with the parameter of the primary also allows representing both cycles in the p–V plane just as was done in the previous chapter with variable buffer pressure; accordingly, the term secondary *trace* would be appropriate. An interesting application of this mode of representation is in the proof of the following theorem.

THEOREM

The mechanical efficiency of a double-acting engine with identical cycles exactly opposite in phase and having a mechanism of constant effectiveness is the same as that of a single-acting engine with the same workspace cycle and having the optimum constant buffer pressure.

The first diagram of Figure 9.6 represents two identical opposing engine cycles; one is the primary workspace cycle and the other is the trace of the secondary. The shaded area is the forced work done on the piston by the combined cycles. The dotted horizontal line through the points of intersection of the cycles divides the relevant areas into two identical but reflected diagrams, one of which is shown in the second part of Figure 9.6. Considering this in isolation, it is the diagram of a single-acting engine with constant buffer pressure. The ratio of the shaded to the interior area is the same for each diagram because of the symmetry. Thus, the mechanical efficiency is the same for the single- and the dual-cycle engine because of the assumed constant mechanism effectiveness and (3.2). Moreover, the buffer pressure line is positioned so that the horizontal base length of the shaded forced work area is the same on the compression as on the expansion side; as explained in Chapter 3, this is exactly the level that is optimum for a constant mechanism effectiveness engine.

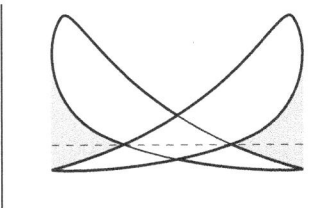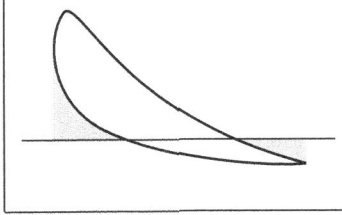

Figure 9.6 An example of *p–V* diagrams for a double-acting engine with identical cycles with 180° phasing.

DOUBLE-ACTING SPLIT-WORKSPACE ENGINES

An example of a more complex engine arrangement is the Siemens-type Stirling engine. Schematically depicted in Figure 9.7, it is a four fold alpha engine formed with four double-acting pistons running 90° apart. The workspaces are interconnected with the upper primary side of each piston being the hot space, and the lower secondary side of the next being the cold space.

Using the same specifications as for the alpha engine example above, the diagram of the force on each piston is shown in Figure 9.8. Numerical integration yields a forced-to-indicated work ratio of 0.43. This is much lower than the 0.81 ratio obtained for the alpha engine with single-acting pistons, but not quite as good as the 0.30 value for the beta Stirling engine of Chapter 3.

With the 0.43 forced-to-indicated work ratio and the same 0.90 mechanism effectiveness, Formula (3.1) or (9.3) gives the following mechanical efficiency limit for the Siemens engine:

$$\eta_m \leq 0.90 - \left(\frac{1}{0.90} - 0.90 \right)(0.43) = 0.809.$$

This is only slightly lower than the 84% mechanical efficiency of the comparable beta Stirling, but much better than the 73% efficiency of the alpha Stirling engine example with single-acting pistons.

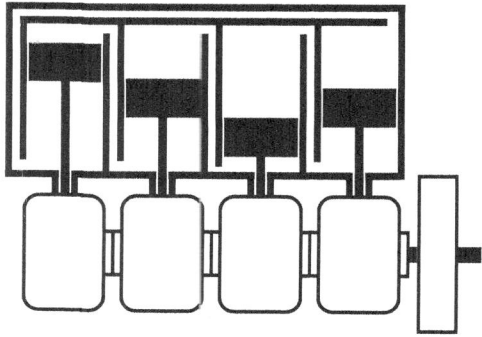

Figure 9.7 The Siemens-type Stirling engine.

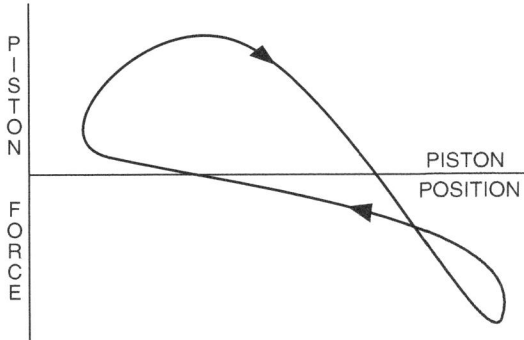

Figure 9.8 Diagram of the force process acting on each piston of a Siemens Stirling engine.

HEAT PUMPS

Figure 9.9 shows an example of a heat pump cycle Λ. In fact, $\Lambda = -\Upsilon$, the reversal of the beta Stirling cycle of Figure 3.1. Here, $W\langle\Lambda\rangle$ numerically equals the area within the cycle but is negative because of the counterclockwise orientation of the heat pump cycle. The shaded area

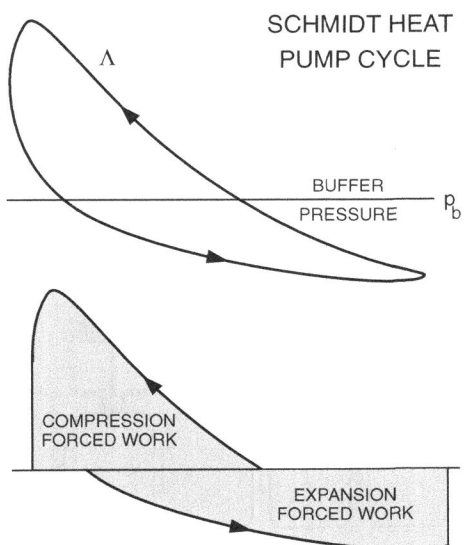

Figure 9.9 A Schmidt heat pump cycle and its forced work.

in the second part of the figure is the forced work $W_-\langle\Lambda\rangle$ for this cycle relative to the constant buffer pressure level shown.

At this point, note that the analysis of Chapter 2 leading up through Inequality (2.3) is valid for heat pump cycles as well as for engine cycles. Net work must be done on the shaft of a heat pump over a cycle if it really pumps heat, so the standard shaft work output W_s as defined for engines is negative for heat pumps. When dealing with pumps, it is perhaps less confusing to work with net shaft input, so the following definition is made:

$$W_{si}\langle\Lambda\rangle = -W_s\langle\Lambda\rangle.$$

This is positive for a heat pump (and negative for an engine!). Now using the right-hand part of Inequality (2.3), some rearrangement gives the following lower bound on the shaft work required to drive a heat pump cycle.

HEAT PUMP THEOREM

If Λ is a heat pump cycle with a constant buffer pressure driven by a mechanism having effectiveness $\varepsilon \leq E$, then the net shaft work input $W_{si}\langle\Lambda\rangle$ required to drive the pump over a cycle is bounded as follows:

$$W_{si}\langle\Lambda\rangle \geq E|W\langle\Lambda\rangle| + \left(\frac{1}{E} - E\right) W_-\langle\Lambda\rangle. \qquad (9.4)$$

If the effectiveness equals the constant E over the whole cycle of operation, then equality holds.

Note that the theorem also would apply to a compressor, or to any pump, if the cycle Λ is taken as that of the fluid pressure against the piston face versus the volume the piston sweeps.

An overall measure of how well a heat pump works is its *mechanical coefficient of performance,* or MCOP. This is defined here as the ratio of thermal energy Q lifted or delivered, according to whether it is utilized as a refrigerator or a heat pump, respectively, to the mechanical shaft input work per cycle:

$$\text{MCOP} = \frac{Q}{W_{si}}.$$

Since the (indicated) COP is defined as $Q/|W|$, the relation between the two coefficients is

$$\text{MCOP} = \frac{|W|}{W_{si}}\text{COP}.$$

This yields the following interesting corollary of the Heat Pump Theorem:

$$\text{MCOP} \leq \frac{\text{COP}}{E + ((1/E) - E)(W_-/|W|)} \tag{9.5}$$

As an example of the application of (9.5), consider again the cycle Λ shown in Figure 9.9. The ratio $W_-/|W| = 1.30$. If $E = 0.90$, then (9.5) shows that the MCOP cannot exceed 85% of the indicated COP:

$$\frac{\text{MCOP}}{\text{COP}} \leq \frac{1}{0.90 + ((1/0.90) - 0.90)(1.30)} = 0.851.$$

10

OPTIMUM STIRLING ENGINE
GEOMETRY

This chapter applies the Fundamental Efficiency Theorem to a central problem in basic Stirling engine design, that of identifying optimal engine geometry. This problem was treated in Chapter 7 for highly idealized engines having theoretical mechanisms, heat exchangers, etc. to produce cycles consisting of four distinct uniform thermodynamic processes. The results in Chapter 7 clearly showed the influence that the type of thermodynamic processes and the level of mechanism effectiveness have on optimum compression ratio and engine output potential.

In this chapter, a more realistic mechanical model of the Stirling engine is employed. It faithfully reflects practical and typical mechanical motions for the piston and displacer. In this setting, optimum values of two parameters are identified which yield maximum brake work output. In the interest of mathematical tractability, the thermal model used here is still highly idealized in that limitations in heat transfer are not considered. Accordingly, it yields best-case results, but allowing for this in a rational way when applying the optima in practical situations can provide an improved guide for first-order design of new engines.

THE GAMMA ENGINE

The analysis is limited here to a particular type of Stirling known as the gamma or split-cylinder. Illustrated in Figure 10.1, the split-cylinder is the simplest of the three main Stirling engine configurations. It is in this

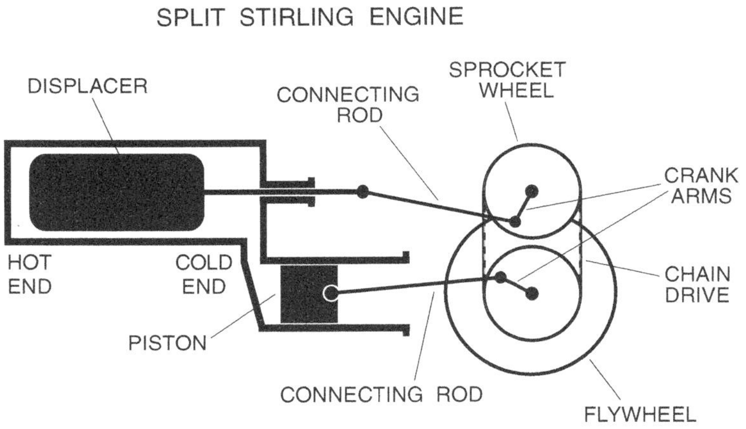

SPLIT STIRLING ENGINE

Figure 10.1 Diagram of the split-cylinder or gamma-type Stirling engine.

type of Stirling that the optimization problem is most familiar and clear. It is the problem that immediately occurs to everyone who sets out to design and make a gamma Stirling engine for the first time. What should be the ratio of the swept volume of the piston to that of the displacer, and what should be the phase angle between them?

No one really worries too much about the phase angle optimum because on many engines it is usually easy to adjust or reset, and so the best phase angle can be experimentally determined to the engine operator's satisfaction. But the swept volume ratio is a much more difficult matter. The ratio cannot be easily changed without affecting other engine features that might themselves affect output, such as overall size and dead volume. An analytical approach is essential not only because of the experimental difficulties, but also because a mathematical basis always directs practical work in a credible direction and guides the interpretation of findings, be they experimental or computational.

THE SCHMIDT ANALYSIS

In an outstanding piece of mathematical analysis, Gustav Schmidt developed a basic model of the Stirling engine and obtained a closed-form expression for its indicated cyclic work (Schmidt, 1871). It is an idealized model, but one which superbly captures essential features of the engine, especially the basic interplay between the mechanically constrained motion of it parts and the resulting thermodynamic cycle. It provides a best-case analysis that at the very least helps identify general Stirling engine characteristics and behavior. Schmidt used it to quantify and interpret the performance potential of contemporary engines.

The form of the Schmidt indicated work formula is far too complex to reveal analytically the optimum geometry of a Stirling engine. In addition, the calculations required to obtain numerical approximations of the optima were too many and too long to do before the advent of the electronic computer. It was about 90 years after Schmidt's pioneering mathematical work that the optimization problem was first taken up (Finkelstein, 1960; Kirkley, 1962; Walker, 1962). The work focused on finding the maximum indicated cyclic work relative to a characteristic cycle pressure and relative to the total mass of the working fluid or to the total swept volume.

As already pointed out in Chapter 7, in light of the Fundamental Efficiency Theorem, it is quite clear that obtaining the maximum indicated output does not ensure getting the maximum shaft or brake output. As has been shown throughout this book, shaft work is not a simple multiple of indicated work but depends upon the shape of the engine cycle and the relative buffer pressure, as well as on the effectiveness of the engine mechanism. In this chapter, the Fundamental Efficiency Theorem is combined with Schmidt's model, and together new optima corresponding to maximum shaft work output are obtained (Senft, 2001, 2002).

THE SCHMIDT MODEL FOR GAMMA ENGINES

The gamma-type Stirling engine can be mathematically described by the following parameters and variables:

V_1 = displacer swept volume
V_2 = piston swept volume

V_D = dead volume

T_H = hot space temperature

T_C = cold space temperature

T_D = dead space temperature = $(T_H + T_C)/2$

$\overline{p}$ = root mean cycle pressure = $\sqrt{p_{max}\, p_{min}}$

p_b = external buffer pressure = $\overline{p}$

ω = angular velocity of crankshaft

α = angle by which displacer crank leads piston crank

τ = T_C/T_H = ratio of temperatures of cold to hot space

κ = V_2/V_1 = ratio of piston swept volume to displacer swept volume

χ = V_D/V_1 = dead volume ratio

ωt = instantaneous angular position of piston crank

V = instantaneous total engine volume

p = instantaneous pressure throughout engine spaces

In spite of the finite-length connecting rods shown in Figure 10.1, a standard assumption of the Schmidt model is pure sinusoidal motion of the piston and displacer. This is a reasonable and tractable approximation of the motions in most real engines. The other Schmidt assumptions include an ideal gas as the working fluid; isothermal hot, cold, and dead spaces; uniform instantaneous pressure throughout all the engine spaces; and no leakage of working gas into or out of the engine.

Convenient expressions for the instantaneous total engine volume V and pressure p are the following:

$$V = \frac{V_T}{\kappa + 1}\left(1 + \frac{\kappa}{2}(1 + \cos(\omega t)) + \chi\right)$$

and   (10.1)

$$p = \overline{p}\,\frac{\sqrt{Y^2 - X^2}}{Y + X\cos(\omega t - \theta)}$$

where

$$X = \sqrt{\kappa^2 - 2\kappa(1 - \tau)\cos\alpha + (1 - \tau)^2}, \quad Y = 1 + \tau + \kappa + \frac{4\tau\chi}{1 + \tau},$$

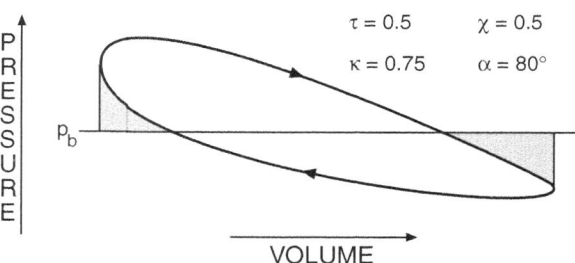

Figure 10.2 The pressure–volume diagram of a gamma-type Stirling engine plotted from the Schmidt model. The shaded area is the forced work.

and

$$\theta = \arccos\left(\frac{\kappa - (1 - \tau)\cos\alpha}{X}\right).$$

Figure 10.2 shows a p–V diagram of a gamma engine based upon Equations (10.1) for some practical parameter values.

INDICATED WORK

A closed-form expression for the indicated work per cycle $W = \oint p\, dV$ of a Schmidt gamma Stirling can be written in the following way:

$$W = \frac{V_T \, \overline{P}}{\kappa + 1} \frac{\pi(1 - \tau)\kappa \sin\alpha}{\sqrt{Y^2 - X^2} + Y} \tag{10.2}$$

where X and Y are the functions defined above in the set of Equations (10.1). Appendix C outlines the derivation of Equations (10.1) and (10.2). Note that the only factors having dimension in the expression for W are the combined or *total swept volume* of the piston and displacer, $V_T = V_1 + V_2 = (\kappa + 1)V_1$, and $\overline{p}$, the root mean of the maximum and minimum cycle pressures, which also equals the integral average pressure. For simplicity in this treatment, all the dead space is treated as being at the arithmetic average of the extreme cycle temperatures; this is reflected in the last term of the parameter Y which involves the dead space ratio χ.

SHAFT WORK

In the analysis in this chapter, mechanism effectiveness will be assumed constant. This is appropriate for first-order-level analysis and will result in a best-case analysis as explained in Chapter 3. Accordingly, cyclic shaft work for this chapter will be taken from Equation (3.2):

$$W_s = E W - \left(\frac{1}{E} - E \right) W_- \tag{3.2}$$

where W is the indicated work of the cycle and W_- is the forced work associated with the cycle and its buffer pressure.

PARAMETER EFFECTS ON BRAKE OUTPUT

The values of the parameters in the Schmidt model influence not only the indicated work according to (10.2), but also the forced work. As an illustration, Figure 10.3 shows a sequence of three Schmidt gamma engine cycles with differing piston-to-displacer swept volume ratios, $\kappa = V_2/V_1$. In each of the cycles, the displacer swept volume V_1 and the mean cycle pressure $\bar{p}$ are the same, and are taken as unity. Each engine cycle has the same temperature ratio $\tau = 0.7$ and phase angle

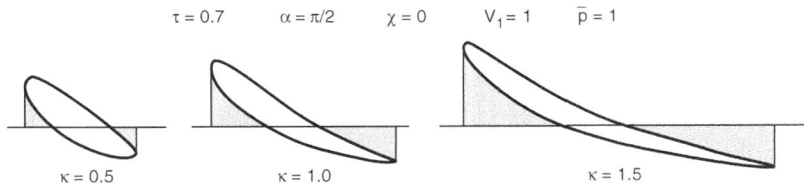

Figure 10.3 The effect of changing the swept volume ratio on the pressure–volume cycle of a Schmidt gamma engine with fixed displacer volume. The table shows calculated work values and mechanical efficiency based on a mechanism effectiveness of $E = 0.8$.

κ	0.5	1.0	1.5
W	.109	.182	.235
W_-	.022	.119	.273
W_s	.077	.092	.066
η_m	.71	.50	.28

$\alpha = \pi/2$, and no dead space $V_D = 0$. The values of the swept volume ratio κ for the three cycles are 0.5, 1.0, and 1.5.

The cycle diagrams are all drawn to the same scale. Because the displacer swept volume is the same for the three cases, the cycles in this example increase in size with κ. The table gives the work values calculated from the Schmidt model. The indicated work W of the largest cycle is over twice that of the smallest. Of special note is how quickly forced work increases. The largest cycle has over ten times the forced work of the smallest, and it exceeds the indicated work that its cycle produces.

The last two lines of the table give the shaft work output W_s and the mechanical efficiency $\eta_m = W_s / W$ of the cycles when coupled to a mechanism of constant effectiveness $E = 0.8$. The shaft work output is largest for the middle cycle and lowest for the largest cycle. Mechanical efficiency is highest for the smallest cycle and decreases as κ increases. This example indicates how significantly the optimum parameters for the Schmidt model can be influenced by taking mechanical friction losses into account.

OPTIMUM SWEPT VOLUME RATIO AND PHASE ANGLE

When making comparisons of engine output, it is only fair to consider engines of the same size. A natural nominal measure of size for gamma Stirling engines is the total combined swept volumes of the piston and displacer. This is the size measure advocated in the earlier optimizations of indicated work (Kirkley, 1962; Walker, 1962). The optimization problem then becomes one of partitioning a given total swept volume between the piston and displacer sections of the engine, i.e., choosing κ, to maximize the work per cycle. The other parameter subject to optimization is the phase angle, α, by which the motion of the displacer leads that of the piston.

A fair comparison also requires that the engines have the same characteristic pressure. The mean cycle pressure $\overline{p}$ is the most natural

and intuitive pressure against which to compare gamma engine performance. It is also the easiest characteristic pressure to maintain and
measure in practice, because it automatically tends, or can easily be
made, to equalize to the buffer or surrounding atmospheric pressure,
as has been already noted in earlier chapters.

Presented in Table 10.1 are the results of numerical calculations
using Equations (10.1), (10.2), and (3.2) for a range of values of
temperature ratio τ, mechanism effectiveness E, and dead space
ratio χ. The table shows the values κ^* and α^* of swept volume ratio
and phase angle that yield the maximum specific shaft work for the
given values of τ, E, and χ. The specific shaft work $\tilde{W}_s$ here is defined
relative to the total swept volume and mean cycle pressure as already
discussed, namely,

$$\tilde{W}_s = \frac{W_s}{\bar{p}\, V_T}.$$

Its maximal values shown in the table are denoted by $\tilde{W}_s^*$. Also shown
in the table are the corresponding values of specific indicated work
$\tilde{W} = W/(\bar{p}V_T)$, specific forced work $\tilde{W}_- = W_-/(\bar{p}V_T)$, and the mechanical efficiency η_m, all computed at the point where the maximum spe-
cific shaft work $\tilde{W}_s^*$ occurs, namely for the parameter values
$\tau, \chi, E, \kappa = \kappa^*$, and $\alpha = \alpha^*$.

SWEPT VOLUME RATIO SELECTION

In applying the optimization Table 10.1, it is important to keep in mind
the interchange between mechanical efficiency and maximum shaft
output. As can be seen from the table, at the maximum shaft output
point, the mechanical efficiency is most often significantly below the
mechanism effectiveness. A better mechanical efficiency can be
obtained by taking a swept volume ratio smaller than κ^*. In most
cases, just a slightly smaller κ will considerably improve mechancal
efficiency while resulting in only a relatively small reduction in shaft
output. This is illustrated in Figure 10.4, which gives a graphical picture

10.1 OPTIMUM SCHMIDT GAMMA ENGINE GEOMETRY								
τ	χ	E	κ^*	α^*	$\tilde{W}_s^*$	$\tilde{W}$	$\tilde{W}_-$	η_m
0.3	0	0.7	0.78	84°	0.157	0.245	0.0199	0.64
0.3	1	0.7	0.89	82	0.107	0.170	0.0166	0.63
0.3	0	0.8	0.90	88	0.189	0.254	0.0318	0.74
0.3	1	0.8	1.05	85	0.130	0.178	0.0271	0.73
0.3	0	0.9	1.06	91	0.224	0.261	0.0482	0.86
0.3	1	0.9	1.28	89	0.157	0.184	0.0450	0.85
0.4	0	0.7	0.74	83°	0.123	0.195	0.0188	0.63
0.4	1	0.7	0.83	81	0.0788	0.128	0.0146	0.62
0.4	0	0.8	0.87	86	0.149	0.204	0.0302	0.73
0.4	1	0.8	1.00	84	0.0973	0.136	0.0250	0.72
0.4	0	0.9	1.06	90	0.179	0.211	0.0499	0.85
0.4	1	0.9	1.28	88	0.119	0.143	0.0450	0.83
0.5	0	0.7	0.68	81°	0.0921	0.149	0.0167	0.62
0.5	1	0.7	0.75	80	0.0565	0.0938	0.0126	0.60
0.5	0	0.8	0.82	85	0.114	0.158	0.0292	0.72
0.5	1	0.8	0.93	84	0.0710	0.102	0.0233	0.70
0.5	0	0.9	1.03	89	0.139	0.166	0.0503	0.84
0.5	1	0.9	1.25	88	0.0893	0.110	0.0451	0.81
0.6	0	0.7	0.60	80°	0.0649	0.108	0.0146	0.60
0.6	1	0.7	0.64	81	0.0382	0.0653	0.0103	0.59
0.6	0	0.8	0.73	84	0.0817	0.117	0.0259	0.70
0.6	1	0.8	0.83	83	0.0492	0.0729	0.0202	0.68
0.6	0	0.9	0.98	88	0.103	0.126	0.0503	0.82
0.6	1	0.9	1.16	87	0.0639	0.0808	0.0418	0.79
0.7	0	0.7	0.49	79°	0.0411	0.0705	0.0113	0.58
0.7	1	0.7	0.53	79	0.0234	0.0415	0.00785	0.56
0.7	0	0.8	0.64	84	0.0532	0.0799	0.0237	0.67
0.7	1	0.8	0.69	83	0.0309	0.0477	0.0160	0.65
0.7	0	0.9	0.89	87	0.0696	0.0885	0.0476	0.79
0.7	1	0.9	1.03	87	0.0420	0.0554	0.0371	0.76
0.8	0	0.7	0.37	80°	0.0212	0.0388	0.00824	0.55
0.8	1	0.7	0.37	78	0.0116	0.0211	0.00433	0.55
0.8	0	0.8	0.49	83	0.0286	0.0453	0.0171	0.63
0.8	1	0.8	0.54	82	0.0160	0.0266	0.0117	0.60
0.8	0	0.9	0.74	87	0.0399	0.0537	0.0403	0.74
0.8	1	0.9	0.83	86	0.0232	0.0325	0.0286	0.71
0.9	0	0.7	0.21	78°	0.00639	0.0127	0.00340	0.50
0.9	1	0.7	0.21	78	0.00336	0.00666	0.00179	0.50
0.9	0	0.8	0.29	82	0.00923	0.0160	0.00800	0.58
0.9	1	0.8	0.30	82	0.00494	0.00878	0.00462	0.56
0.9	0	0.9	0.50	86	0.0145	0.0220	0.0250	0.66
0.9	1	0.9	0.53	86	0.00806	0.0126	0.0155	0.64

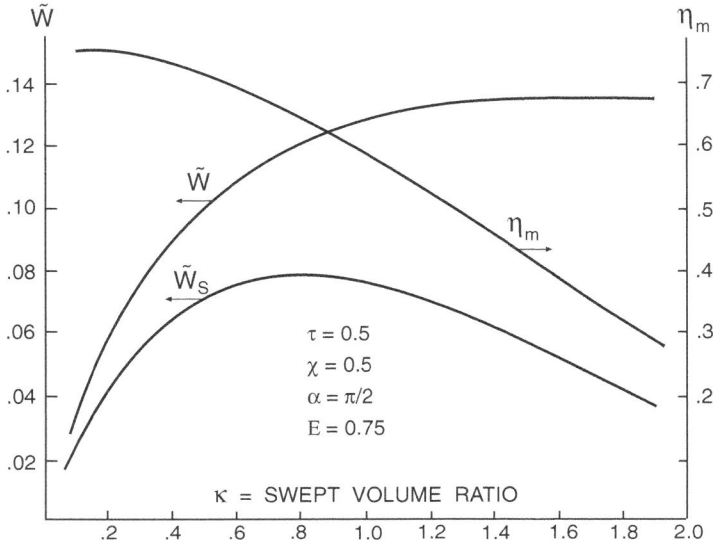

Figure 10.4 An example of the variation of specific indicated work, specific shaft work, and cyclic mechanical efficiency with respect to swept volume ratio for a gamma-type Schmidt Stirling engine.

of how the swept volume ratio affects specific indicated work, specific shaft work, and mechanical efficiency for the case of an engine with parameter values $\tau = 0.5$, $\chi = 0.5$, $E = 0.75$, and $\alpha = \pi/2$.

The specific works for the example of Figure 10.4 are calculated just as in the optimization, that is, relative to total swept volume and mean cycle pressure. Although the peak specific shaft output occurs at a swept volume ratio just below 0.8, there is not much reduction in shaft output with $\kappa = 0.6$ or even 0.5, and mechanical efficiency is much improved. At $\kappa = 0.8$, specific shaft work $\tilde{W}_s = 0.077$ and mechanical efficency $\eta_m = 0.64$, whereas at $\kappa = 0.5$, $\tilde{W}_s = 0.071$ and $\eta_m = 0.71$. In this case, the small reduction in shaft output could be well offset by the much improved mechanical efficiency, which translates to less wear and longer life.[†] On the other hand, the graphs show

[†] The parameters in this example were taken from the author's 10" fan Moriya (Senft, 1974), which was built with a swept volume ratio of 0.56.

that any value of κ larger than 0.8 would lower both output and mechanical efficiency.

The general lesson from such examples is that *it is always advisable to choose a swept volume ratio smaller than the shaft work optimum* shown in the table. How much smaller is a matter of careful compromise between shaft work and mechanical efficiency. The work output curves in Figure 10.4 drop off rather rapidly as κ is taken farther below κ^*. A very small choice for κ would give good mechanical efficiency, but the engine output could be undesirably low. Thus, the choice of κ should be lower, but not too much lower, than κ^*. In a specific case, a plot of $\tilde{W}_s$ should be made as in Figure 10.4, calculating from the equations given above, and from this, the best choice can be made.

If it is desired that the engine perform well over a range of temperature ratios, plots can be made for several values of τ spanning the range, and all the plots taken into account when making the final selection of κ. There could be other considerations to take into account as well. For example, a smaller κ would allow the engine to start at a lower hot end temperature, which could be important say for an engine that drives a pump to pressurize itself. Most important however, is that limited heat transfer rates strongly favor choosing a smaller κ.

INTERNAL TEMPERATURES

It is essential to keep in mind that the analysis here is based on the premise that the temperature ratio represents the temperature extremes of the working gas. Because of heat transfer limitations, the gas temperatures will always be closer together than the temperatures of the metal surfaces of the heat exchangers. This makes the actual gas temperature ratio τ larger than the ratio of the cold-to-hot end surface temperatures. This temperature ratio discrepancy becomes more pronounced at higher engine speeds. This needs to be taken into account when choosing the swept volume ratio for an engine design. The optimization table shows that as the effective gas temperature ratio τ increases, the optimal swept volume ratios are lower. Once again the general advice that emerges is to *favor a swept volume ratio lower than* κ^*. Again, how much lower depends upon how effective the heat exchangers are and how fast the engine will be run.

INDICATED WORK MAXIMA

The analysis done here further shows that the point of maximum indicated work is quite far from the maximum shaft work point and is in fact on the unfavorable side of the brake output curve. In the case shown in Figure 10.4, the maximum indicated work $\tilde{W} = 0.135$ occurs at about $\kappa = 1.7$ and $\alpha = 90°$. At this point, shaft work is a low $\tilde{W}_s = 0.048$ and mechanical efficiency is a dismal $\eta_m = 0.35$.

As another example of this, extreme but real, consider the design of an engine intended to operate at a low temperature differential with $\tau = 0.9$, and, for simplicity, assume $\chi = 0$. Numerical optimization shows that the maximum specific indicated work is $\tilde{W} = 0.0292$, and this is attained when $\kappa = 1.6$ and $\alpha = 89°$. At this point, $\tilde{W}_- = 0.135$. Even if the mechanism effectiveness is a fairly good $E = 0.8$, Formula (3.2) shows that shaft work for this engine will be negative:

$$\tilde{W}_s = E\tilde{W} - (1/E - E)\tilde{W}_- =$$
$$= (0.8)(0.0292) - (1/0.8 - 0.8)(0.135) = -0.0374 .$$

In other words, this engine could not even run itself. Even if the mechanism effectiveness were raised to 0.9, the engine still could not run itself. From the optimization table 10.1, maximum specific shaft work for a Schmidt gamma engine with $\tau = 0.9$, $\chi = 0$, and $E = 0.8$ occurs at a quite different geometry, namely, $\kappa^* = 0.29$. This is a drastically lower value than $\kappa = 1.6$ for the indicated work maximum. Moreover, by the discussion above, a slightly lower value still should be chosen in practice to improve mechanical efficiency. This geometry is consistent with actual low temperature differential Stirling engine experience (Senft, 1986).

As already noted, the most natural and practical characteristic pressure to use when comparing gamma-type Stirling engines is the mean cycle pressure. However, the maximum cycle pressure appeared as another possible choice among some early researchers working with maximum indicated work. When this characteristic pressure was tried in the brake work optimization, sample calculations showed that the optimum swept volume ratios tended to be lower than when mean cycle pressure was used. But even optimizing indicated cyclic work per unit maximum cycle pressure yields engine geometries that are not practical. In the above example with $\tau = 0.9$ and $\chi = 0$, the maximum indicated work per unit maximum cycle pressure occurs at $\kappa = 0.9$. This is still very much larger than $\kappa^* = 0.29$ and also yields an engine which will not run if its mechanism effectiveness is 0.8 or less.

PHASE ANGLE

Examination of the optimization table reveals that the optimum phase angle α^* tends to be lower for poorer mechanisms and for higher temperature ratios. Although 90° is usually considered near enough to the optimum for all practical purposes in gamma engines, the table shows values lower by more than 10° in some cases. Of course, in most split-cylinder engines, the phase angle can be easily changed to best suit operating conditions.

On the other hand, engine performance is not extremely sensitive to phase angle. Figure 10.5 shows how the engine of Figure 10.4 responds to variations in its phase angle. The swept volume ratio for this example was taken to be $\kappa = 0.75$, for which chosen value the best phase angle is near 80°. The table in the figure gives the calculated specific work values in a neighborhood of this point; it shows that shaft output is virtually the same for 10° on either side of the optimum 80°. Note that the maximum indicated work in this example occurs at 90°, for the fixed swept volume ratio.

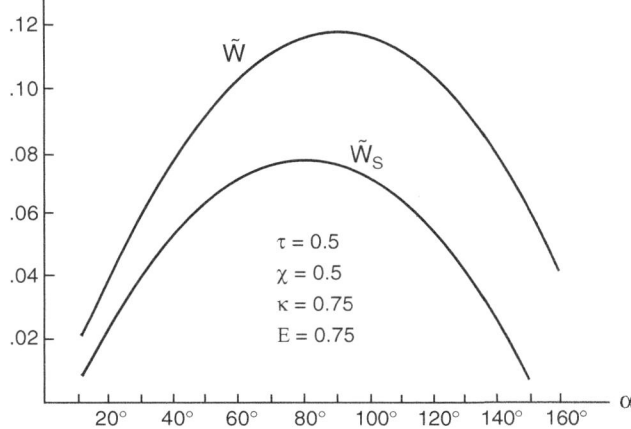

Figure 10.5 Variation of specific indicated and shaft work with phase angle for the gamma engine of Figure 10.4 with a fixed swept volume ratio of 0.75.

α	50°	60°	70°	80°	90°	100°	110°
$\tilde{W}$	0.089	.101	.110	.116	.118	.117	.112
$\tilde{W}_-$	0.008	.010	.012	.016	.020	.025	.031
$\tilde{W}_s$	0.062	.070	.076	.078	.077	.073	.066
η_m	0.70	.69	.68	.67	.65	.63	.59

DEAD SPACE EFFECTS

The negative effect of dead volume on the indicated work of the Stirling cycle is well known. Shaft work is likewise affected. For example, from the optimization Table 10.1, a high performance engine with $\tau = 0.3$ and $E = 0.8$ has a maximum specific brake output of 0.189 with no dead space, and only 0.130 with $\chi = 1$; this amount of dead space reduces the output potential by about 31%. It is important to keep this in mind when designing internal heat exchangers to improve Stirling engine performance. The increase in exchanger surface area must be large enough to more than offset in increased speed the decrease in cyclic work caused by the dead volume of the heat exchanger passages.

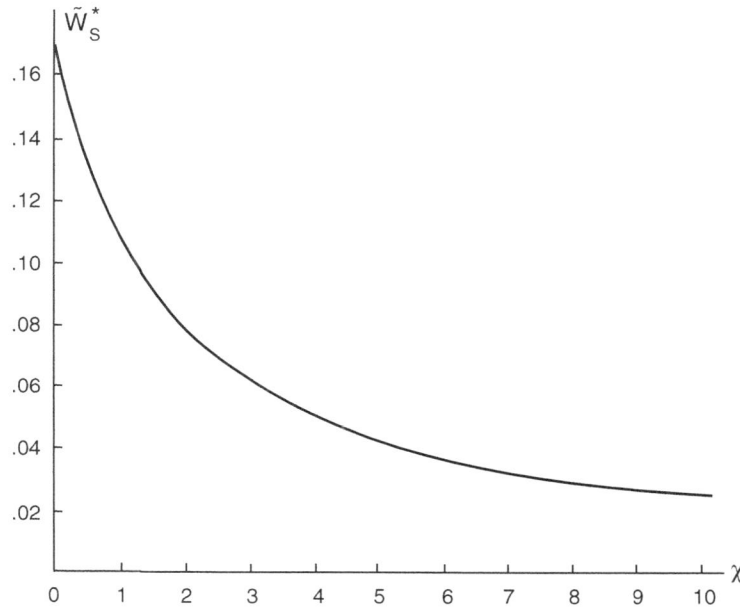

Figure 10.6 Variation of maximum specific shaft work $\tilde{W}_s^*$ versus dead space ratio χ for Schmidt gamma engines with temperature ratio $\tau = 0.5$ and a mechanism effectiveness of $E = 0.7$. The optimum swept volume ratio κ^* ranges from 0.96 to 1.14, and the optimum phase angle α^* ranges from 80° to 78°.

The dead volume effect is greater for engines operating on smaller temperature differences as is seen, for example, from the Table 10.1 entries for $\tau = 0.8$ and $E = 0.8$. Here the maximum shaft work with no dead space is 0.0286, while $\chi = 1$ lowers it to 0.0160, a reduction of 44%.

Figure 10.6 shows the variation of specific shaft work with increasing dead space over the range from $\chi = 0$ to 10 for a particular engine with $\tau = 0.5$ and $E = 0.7$. Note that performance drops off at a high rate over the entire usual range of relative dead volume, becoming what one might call gradual only for dead volume ratios much higher than ever necessary in practice.

Another point worth observing from the optimization table is that the optimum swept volume ratio increases as dead volume is increased. However, the increase is relatively small, and, most importantly, dead volume effects on brake output are not neutralized by increasing compression ratio, as one might have guessed prior to the analysis.

ALTERNATE ENGINE CONFIGURATIONS

The analysis presented here can be applied to the single-cylinder or beta-type Stirling in much the same way as for the gamma. The Schmidt equations need to be modified only slightly to reflect overlapping of the piston and displacer strokes. In the case of the alpha or two-piston Stirling engine, the shaft work Formula (3.2) must be applied to each piston and then combined as described in Chapter 9 for split-workspace engines.

CONCLUSIONS

The analysis in this chapter shows again that optimum engine parameters are not solely thermodynamically determined, but are also heavily dependent upon the mechanical section of the engine. Indicated work optima do not occur at the same point as brake work optima, and indeed some indicated work optima yield engines that cannot run in spite of having quite good mechanisms. Some of the findings important for engine design may be summarized as follows:

- Maximum shaft output occurs at smaller swept volume ratios than does maximum indicated work.
- Less effective mechanisms favor smaller swept volume ratios.
- Smaller swept volume ratios yield better mechanical efficiency.
- Low temperature differential engines require small swept volume ratios.
- Dead volume incurs a high penalty in brake output.
- Higher engine speeds favor lower swept volume ratios.
- Dead volume effects cannot be offset by increasing compression ratio.

11

HEAT TRANSFER EFFECTS

This chapter continues the examination of the limits on Stirling engine performance by taking into consideration, with the mechanical losses already covered, thermal limitations and losses from which real Stirling engines suffer. First covered is limited heat transfer rate into and out of the working fluid of the engine. This is modeled here just as Curzon and Ahlborn did for Carnot engines (Curzon & Ahlborn, 1975). In addition, introduced later in the chapter is an internal heat leak through the engine from the hot to the cold section governed by the same heat transfer regime. This simulates in a general way the various internal thermal losses occurring in real Stirling engines.

HEAT EXCHANGE

Thermal energy must be transferred into and out of a Stirling engine via heat exchangers at the hot and cold ends. A temperature gradient is required to drive the transfer; in other words, there must be a temperature differential between the source reservoir and the working fluid when it receives thermal energy. Likewise, a temperature difference is required between the engine working substance and the sink reservoir in order for the engine to reject thermal energy. The larger these differences, the greater the rate of energy transfer. This aspect of heat transfer is modeled in a general way by Newton's Law of Cooling (Bejan, 1996b).

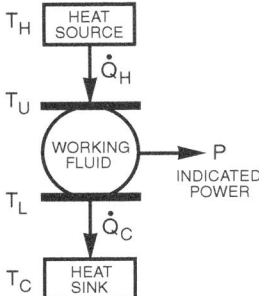

Figure 11.1 Diagram of an engine with finite heat transfer rates.

HEAT TRANSFER ASSUMPTIONS

The thermal features of the general engine model considered here are shown in Figure 11.1. Assume given reservoirs at temperatures T_H and T_C and heat transfer coefficients a and b which govern the flow of thermal energy into and out of the working section of the engine according to Newton's law:

$$\dot{Q}_H = a(T_H - T_U) \quad \text{and} \quad \dot{Q}_C = b(T_L - T_C) \tag{11.1}$$

where $T_H \geq T_U \geq T_L \geq T_C$. The temperatures T_U and T_L are those of the working fluid at which it receives and rejects thermal energy with the source and sink reservoirs, respectively. The heat transfer coefficients a and b are characteristics of the particular engine under consideration. It is assumed that the heat exchanges take place for some fixed fractions of the engine cycle period, for example, during the expansion and compression portions of an ideal Stirling cycle as suggested in Figure 11.2. These fractions are taken as solely determined by the engine hardware or characteristics and are essentially independent of the operating frequency of the engine. The constant fractions are taken to be incorporated in the coefficients a and b in Equations (11.1). Thus, $\dot{Q}_H$ and $\dot{Q}_C$ are cycle-time-averaged heat transfer rates. For example, the product of the period of the cycle and $\dot{Q}_H$ is the thermal energy transferred from the source to the engine working fluid for one cycle.

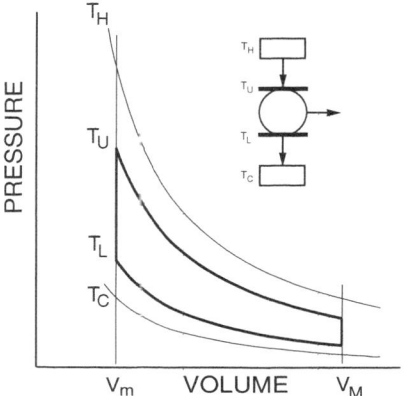

Figure 11.2 An ideal Stirling cycle with limited heat transfer between the source and sink reservoirs.

The *ratio of engine operating temperatures*, lower to upper, is here denoted as before by τ:

$$\tau = \frac{T_L}{T_U}.$$

However, the *reservoir temperature ratio*, colder to hotter, will be now denoted by Γ:

$$\Gamma = \frac{T_C}{T_H}.$$

The *ratio of the upper engine temperature to the hot reservoir temperature* will be denoted by ξ:

$$\xi = \frac{T_U}{T_H}.$$

The following heat transfer coefficient ratio will be convenient:

$$\delta = \frac{b}{a}.$$

For any engine operating subject to the assumptions above, the *average indicated cycle power P* is given by

$$P = \dot{Q}_H - \dot{Q}_C = a(T_H - T_U) - b(T_L - T_C) \tag{11.2}$$

which can be expressed as

$$P = a\,T_H\,[1 + \delta\Gamma - \xi - \xi\delta\tau]. \tag{11.3}$$

Whatever cycle the working fluid undergoes, the Second Law requires that its thermal efficiency not exceed the Carnot efficiency based upon the available temperature extremes; that is, using (11.3) and (11.1),

$$\frac{\dot{Q}_H - \dot{Q}_C}{\dot{Q}_H} = \frac{P}{\dot{Q}_H} \leq 1 - \frac{T_L}{T_U} = 1 - \tau. \tag{11.4}$$

This, which will be referred to as the *Carnot Condition*, can be written as

$$\xi(\delta + 1)\tau \geq \delta\Gamma + \tau. \tag{11.5}$$

MAXIMUM INDICATED POWER

Equality holds in (11.4) and (11.5) for the so-called *endoreversible engines*, that is, those engines having cycles that are internally reversible such as the Carnot or the ideal regenerative Stirling, the latter being the subject of particular interest here. Substituting (11.5) *with equality* into (11.3) gives the following expression for the *cycle-average indicated power* of an endoreversible engine:

$$P_i = \frac{a\,T_H\,\delta}{1 + \delta}\,\frac{(1 - \tau)(\tau - \Gamma)}{\tau}. \tag{11.6}$$

As engine speed varies over its permissible range, the engine operating temperature ratio τ ranges from Γ to 1. At each extreme, $P_i = 0$. By differentiating (11.6) and solving for zero, it is easily found that the maximum indicated power P_i is attained at the speed where the engine operating temperature ratio τ is $\sqrt{\Gamma}$. Thus, the thermal efficiency of an endoreversible engine operating at its maximum indicated power level is $1 - \sqrt{\Gamma}$. This is the famous result first obtained by Curzon and Ahlborn (1975).

MAXIMUM BRAKE POWER

For given values of a, T_H, δ, Γ, and τ, the *shaft power* realized by an ideal Stirling is the product of indicated power (11.6) and mechanical efficiency (3.6):

$$P_s = P_i\,\eta_{ms}(\mathrm{E}, \tau, r). \tag{11.7}$$

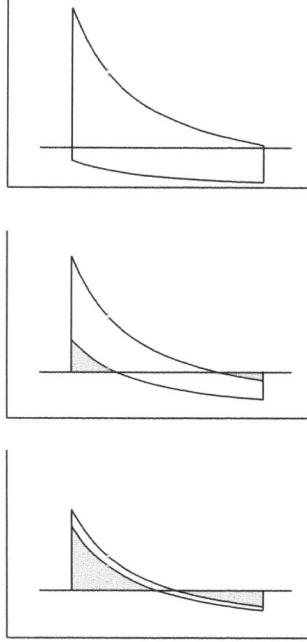

Figure 11.3 As engine speed increases with limited heat transfer, the Stirling cycle becomes thinner, forced work increases, and mechanical efficiency decreases.

Theorem (3.6) showed that $\eta_{ms}(E, \tau, r)$ is the maximum mechanical efficiency for any engine with the same E, τ, and r values. The maximum shaft power P_s with respect to τ, however, will not necessarily occur at the Curzon–Ahlborn (C&A) temperature ratio. Figure 11.3 shows the cycles of a Stirling engine with limited heat transfer operating at pro-gressively higher speeds. At low speeds, the engine operating tempera-tures T_U and T_L are close to the reservoir temperatures T_H and T_C, and mechanical efficiency is high. As speed increases, the engine operating temperatures move away from the reservoir temperatures because of the limited heat transfer. As the cycle becomes thinner, the associated forced work increases, as shown by the shaded area. This causes mechan-ical efficiency to decrease, and the net result is that the maximum brake power point occurs at τ values closer to the reservoir ratio Γ.

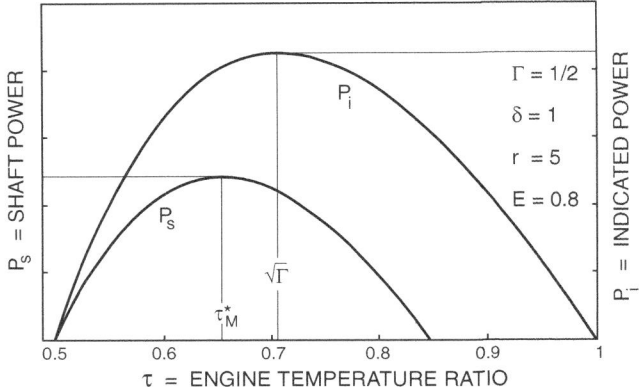

Figure 11.4 Plots of the indicated power and shaft power of an ideal Stirling engine with limited heat transfer as functions of engine operating temperature ratio.

An example of this is shown here in Figure 11.4. Plots of indicated power P_i given by (11.6) and shaft power P_s given by (11.7) are shown for an ideal Stirling engine with reservoir temperature ratio $\Gamma = 0.5$, mechanism effectiveness $E = 0.8$, compression ratio $r = 5$, and heat transfer ratio $\delta = 1$. It can be seen that the *maximum shaft power point*, which will be denoted by τ_M^*, is indeed between the reservoir ratio Γ and the C&A point $\sqrt{\Gamma}$, that is,

$$\Gamma \leq \tau_M^* \leq \sqrt{\Gamma}. \tag{11.8}$$

That (11.8) always holds will be proven in a more general context later in the chapter. This means that the maximum shaft power is attained at the same or a lower engine speed than its maximum indicated power.

Figure 11.5 shows another example with a slightly higher compression ratio $r = 6$ and a very poor mechanism effectiveness of $E = 0.6$. This is an extreme example, but it is interesting because the engine it depicts will not even run itself at its point of maximum indicated power.

However, it is worth noting a condition under which the maximum brake power will occur exactly at the C&A point. In the efficacious range, that is, when $\tau r \leq 1$, $\eta_{ms}(E, \tau, r) = E$. If the compression ratio of the engine happens to be such that $\sqrt{\Gamma} r \leq 1$, then $\eta_{ms} = E$ for τ in

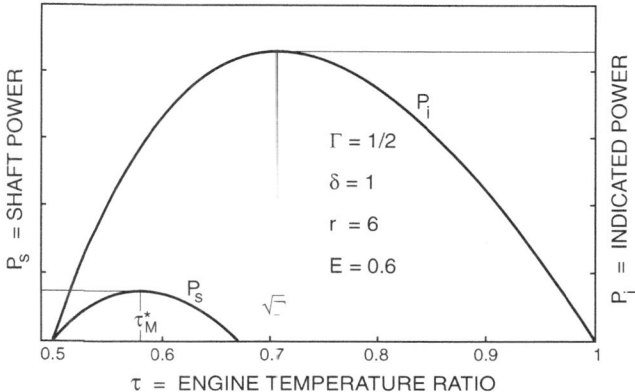

Figure 11.5 Power plots as in Figure 11.4 but with a higher compression ratio and a lower mechanism effectiveness.

the interval $\Gamma \leq \tau \leq \sqrt{\Gamma}$. Because $\eta_{ms} \leq E$ for τ beyond this, that is, when $\sqrt{\Gamma} < \tau \leq 1$, it follows that $\sqrt{\Gamma}$ is where the shaft power is a maximum. This proves the following result.

THEOREM

The shaft power maximum of an ideal Stirling engine with $r \leq 1/\sqrt{\Gamma}$ and limited heat transfer occurs at the C&A point, $\sqrt{\Gamma}$.

Note that choosing a compression ratio below $\sqrt{1/\Gamma}$ yields an engine that runs efficaciously at its maximum power point, that is, with the best possible mechanical efficiency. In practical terms, this is usually a very low compression ratio. For example, if $\Gamma = 0.25$, $\sqrt{1/\Gamma} = 2$. In fact, actual high performance Stirling engines have low compression ratios.

BRAKE THERMAL EFFICIENCY AT MAXIMUM POWER

Brake thermal efficiency is the product of thermal efficiency and mechanical efficiency. At this point, it is instructive to compare brake

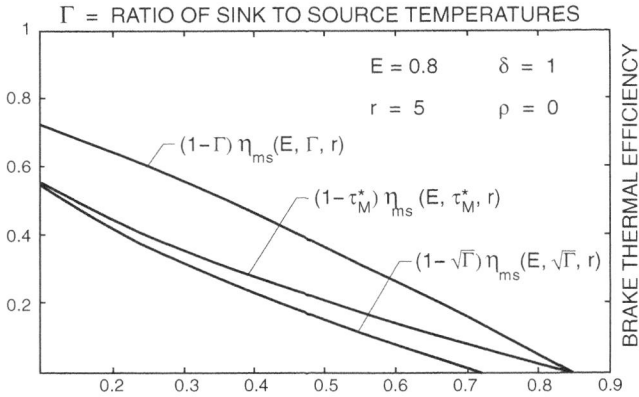

Figure 11.6 Plots of three brake thermal efficiencies versus reservoir temperature ratio.

thermal efficiencies under various conditions. Figure 11.6 shows three plots of brake thermal efficiencies at different operating points as functions of the reservoir temperature ratio Γ for an ideal Stirling engine with compression ratio $r = 5$ and a mechanism of effectiveness $E = 0.8$.

The top curve is the brake thermal efficiency based on the reservoir temperatures. It is the product of the ideal Stirling or Carnot thermal efficiency and the maximal mechanical efficiency η_{ms} at the reservoir temperature ratio Γ:

$$(1 - \Gamma)\,\eta_{ms}(E, \Gamma, r).$$

This efficiency corresponds to the theoretical case of infinite heat transfer rates.

The bottom curve is the brake thermal efficiency at the point of maximum *indicated* power, which is at the (C&A) temperature ratio $\sqrt{\Gamma}$. It is the product of the Carnot efficiency and the maximal mechanical efficiency both evaluated at this point:

$$\left(1 - \sqrt{\Gamma}\,\right)\eta_{ms}(E, \sqrt{\Gamma}, r).$$

This is the efficiency examined by Reader (1991). As seen above, this point is not necessarily the same as the point of maximum shaft power output.

The middle curve gives the true maximum shaft power point efficiency. It is the brake thermal efficiency at the point of maximum shaft power. In other words, it is the product of the Carnot thermal efficiency and the maximal mechanical efficiency η_{ms}, both evaluated at the point τ_M^*, where the shaft power is the maximum:

$$(1 - \tau_M^*) \, \eta_{ms}(E, \tau_M^*, r).$$

The plot of this shown in Figure 11.6 is the case where $\delta = 1$.

It is important to observe that there are noteworthy differences among these efficiency curves in the midrange, which is where most practical Stirling engines operate. The upper two efficiency curves will always meet at zero at the Γ value where $\eta_{ms}(E, \Gamma, r) = 0$ because of (11.8) and the fact that η_{ms} is a decreasing function of temperature ratio, as will be noted. At the other end, when $0 \leq \Gamma \leq 1/r^2$, the lower two curves will coincide because of the previous theorem; for the case shown in Figure 11.6, this occurs in a very small interval outside the range included in the illustration.

HEAT LOSSES IN STIRLING ENGINES

In a working Stirling engine, the hot-end heat exchanger is continually maintained at a high temperature. The cold-end heat exchanger is maintained at a lower temperature. This sets up a temperature gradient within the structure of the engine and its internal components that unavoidably results in the direct flow of some thermal energy from the hot end to the colder end without any intermediate conversion to work. This kind of energy loss will be referred to here as *internal heat leakage*. In a typical real Stirling engine, this leakage is a result of many actions, including simple conduction through the structure connecting the hot section to the cold, through the working substance itself, and through other components such as the displacer and regenerator. Internal thermal losses also appear in more subtle forms, such as so-called *shuttle heat transfer* which is due to the relative motion of engine parts with differing temperatures in near proximity (West, 1986).

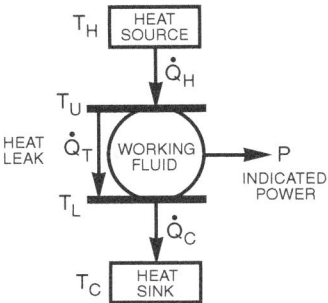

Figure 11.7 Diagram of a heat engine with internal heat leakage.

The combination of all these losses can be represented in a general way as a single heat flow having a rate which is proportional to the difference between the temperature extremes within the engine itself. Figure 11.7 illustrates this heat leakage. This approach captures the overall correlation of internal thermal loss with the engine temperature extremes and provides a suitable point of vantage for the study of its general effect. This is a loss of a different character than the constant external heat leak loss directly between the reservoirs outside of the engine, which has been extensively treated in the literature (Gordon & Huleihil, 1992; Bejan, 1996a,b). An external loss between the reservoirs can easily be added to the analysis presented here if desired.

The heat leak through the engine is assumed to take place between the two internal temperature extremes T_U and T_L at the rate

$$\dot{Q}_T = c(T_U - T_L). \tag{11.9}$$

Again this rate is taken not as an instantaneous value but rather is the cycle-time-averaged heat transfer rate.

The general expression for average *indicated cycle power* with this internal heat leakage turns out to be just as before in (11.2):

$$P = (\dot{Q}_H - \dot{Q}_T) - (\dot{Q}_C - \dot{Q}_T)$$
$$= \dot{Q}_H - \dot{Q}_C = a(T_H - T_U) - b(T_L - T_C).$$

Therefore, (11.3) is formally valid in the present setting. However, the heat leakage does affect the engine operating temperatures T_U and T_L relative to the reservoirs and therefore also the value of the parameter ξ in (11.3). In place of (11.4), the Second Law requirement for an engine with internal heat leakage takes the form

$$\frac{\dot{Q}_H - \dot{Q}_C}{\dot{Q}_H - \dot{Q}_T} = \frac{P}{\dot{Q}_H - \dot{Q}_T} \leq 1 - \frac{T_L}{T_U} = 1 - \tau$$

and therefore the *Carnot Condition* is now

$$\xi\left[(\delta + 1)\tau - \rho(1 - \tau)^2\right] \geq \delta\Gamma + \tau \qquad (11.10)$$

where ρ is the parameter representing the *relative heat leak rate*

$$\rho = \frac{c}{a}.$$

Inequality (11.10) reduces to (11.5) when $\rho = 0$, of course. Equality holds in (11.10) for any endoreversible engine.

Exactly how the engine operating temperatures are related to the reservoir temperatures with heat leakage present can be seen from the Carnot Condition (11.10) together with the requirement that engine power (11.3) be non-negative. Figure 11.8 shows the region of permissible (τ, ξ) points for all possible engines when $\Gamma = \frac{1}{4}$, $\delta = 1$,

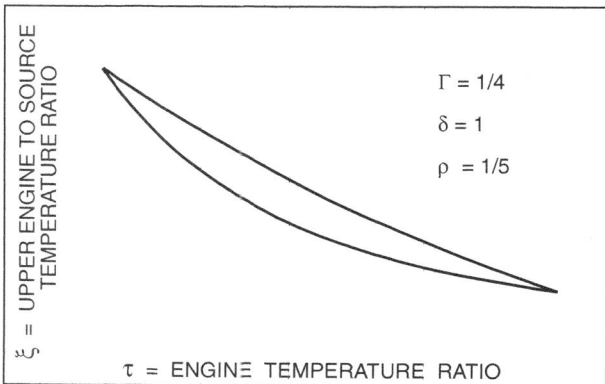

Figure 11.8 Region of possible (τ, ξ) values for engines operating under given heat transfer and leakage parameters.

and $\rho = \frac{1}{5}$. It is easily verified from (11.3) and (11.10) that this illustration is qualitatively representative of the (τ, ξ) domain for all values of the parameters Γ, δ, and ρ. That is, all possible engine cycles operating between given reservoirs and governed by Equations (11.1) and (11.9) are represented by points in a domain having this general shape.

MAXIMUM INDICATED POWER WITH HEAT LEAKAGE

Equation (11.3) shows that the level power curves in order of increasing power form a descending family of hyperbolas "parallel" to the upper boundary curve of the (τ, ξ) region shown in Figure 11.8. Hence, in seeking the maximum possible power, it suffices to consider only the points of the lower boundary curve, which is where equality holds in (11.10), that is:

$$\xi = \frac{\delta\Gamma + \tau}{(\delta + 1)\tau - \rho(1 - \tau)^2}. \tag{11.11}$$

Again, these points correspond to the endoreversible engines. Substituting (11.11) into (11.3) and simplifying gives the formula for their indicated power:

$$P_i = a\, T_H \frac{\delta(\tau - \Gamma)(1 - \tau) - \rho(1 + \delta\Gamma)(1 - \tau)^2}{(\delta + 1)\tau - \rho(1 - \tau)^2}. \tag{11.12}$$

Note that (11.12) gives the average indicated power output of any reversible engine operating between temperatures in the ratio τ for given values of a, T_H, Γ, δ, and ρ. The value of τ yielding the largest value for (11.12) is the maximum indicated power point for the given parameters.

The domain of possible τ values for the indicated power function P_i is the interval from τ_p to 1 where

$$\tau_p = \frac{\delta\Gamma + \rho(\delta\Gamma + 1)}{\delta + \rho(\delta\Gamma + 1)}. \tag{11.13}$$

Indicated power is zero at the endpoints of this interval. Differentiating (11.12) and solving for zero inside the interval yields the following point at which indicated power is maximum:

$$\tau_M =$$

$$\frac{\delta\rho(1-\Gamma) + \sqrt{(1+\delta)(\delta+\rho+\rho\delta)(\rho+\rho\delta^2\Gamma^2+\delta\Gamma(1+\delta+2\rho))}}{\delta+\delta^2+\rho+2\delta\rho+\Gamma\delta^2\rho}$$

This expression for the *maximum indicated power point* τ_M is easily seen to reduce to $\tau_M = \sqrt{\Gamma}$ when $\rho = 0$, giving the result of Curzon and Ahlborn (1975) as a special case.

The complexity of the expression above for τ_M hinders the acquisition of further analytic insights. However, some properties of the graphs of indicated power P_i as given in (11.12) can be easily obtained using elementary calculus. First, for any given values of ρ, δ, and Γ, P_i is a concave downward function of τ in the interval $[\tau_p, 1]$. Second, for any fixed δ, Γ, and τ, P_i is a decreasing function of ρ, and τ_p is an increasing function of ρ. These observations show that the graphs of P_i are always related as depicted in Figure 11.9.

This indicates that τ_M is an increasing function of ρ. In other words, an increase in the internal heat leak moves the maximum indicated

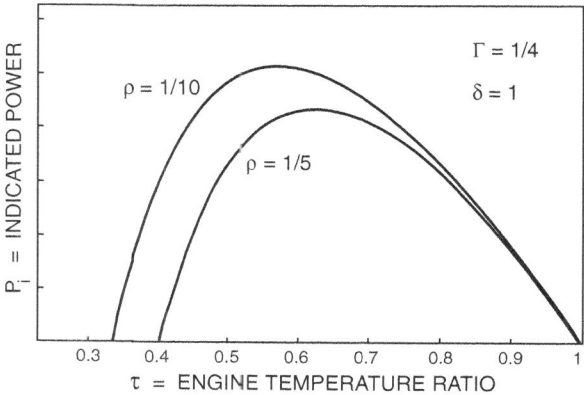

Figure 11.9 Plots of indicated power of an endoreversible engine with different heat leak rates.

power point away from the reservoir temperature ratio. It follows that τ_M is always greater than the Curzon-Ahlborn point $\sqrt{\Gamma}$ when there is heat leakage, that is, when $\rho > 0$. Note that the results thus far apply to all reversible engine cycles, of which the ideal regenerative Stirling is a particular instance, but the one of most interest because of its having the maximal mechanical efficiency, as established in the earlier chapters.

OPERATING FREQUENCY AND TEMPERATURE RATIO IN STIRLING ENGINES

The cycle of an ideal regenerative Stirling engine with the internal heat leak specified above also behaves as shown in Figure 11.3. In other words, with the heat leak added, the ideal Stirling cycle becomes thinner as engine speed increases, just as described earlier without the internal heat leak. With the heat leak, perhaps this is not as intuitively obvious but it can be easily verified analytically.

The exact relationship between *engine frequency f* and the ratio τ of engine operating temperatures for the ideal Stirling cycle can be obtained from the basic relationship $P = fW$, where W is the indicated work per cycle. Equation (4.2) gives W for an ideal Stirling cycle, which here takes the form

$$W = mR(T_U - T_L)\ln r = mRT_H\xi(1 - \tau)\ln r.$$

Using this in the basic relationship together with (11.11) and (11.12) eventually gives

$$f = \frac{a}{mR\ln r}\frac{\delta(\tau - \Gamma) - \rho(1 + \delta\Gamma)(1 - \tau)}{\delta\Gamma + \tau}. \qquad (11.14)$$

This also is rather complex, but elementary calculus shows that the derivative of f with respect to τ is positive for all values of the heat leak coefficient ρ. Hence, engine frequency and engine temperature ratio τ increase or decrease together. So as engine speed increases, the engine operating temperatures T_U and T_L will approach each other, and the engine cycle will indeed become thinner. Therefore, with the heat leak, as without, mechanical efficiency is negatively impacted by increased engine speed. The effect on brake power of ideal Stirlings is examined in the next section.

MAXIMUM BRAKE POWER OF STIRLING ENGINES WITH HEAT LOSS

With internal heat leakage taken into account, the product $P_i \, \eta_{ms}$ of (11.12) and (3.6) gives the highest shaft power output of an ideal Stirling engine with parameter values a, T_H, Γ, δ, ρ, r, E, and τ. Its maximum with respect to τ is thus the absolute maximum shaft or brake power obtainable by a Stirling engine with given parameter values a, T_H, Γ, δ, ρ, r, and E. The closed form for this maximum, if there is one, must be a complex expression. However, graphs and some elementary analysis can again give some interesting insights.

First, consider how η_{ms} varies with τ for fixed values of E and r. Figure 11.10 illustrates a typical case. From Equations (3.5) and (3.6), η_{ms} will be constant and equal to E for τ in the interval $(0, 1/r]$. Beyond this interval, elementary calculus verifies that η_{ms} will always monotonically decrease and become zero at some $\tau < 1$, for all permissible values of E and r. Thus, Figure 11.10 shows typical characteristics of η_{ms} as a function of τ.

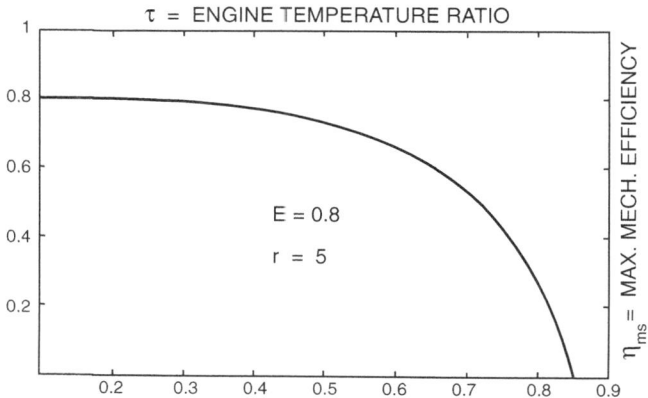

Figure 11.10 $\eta_{ms}(E, \tau, r)$ plotted as a function of τ for fixed values of E and r.

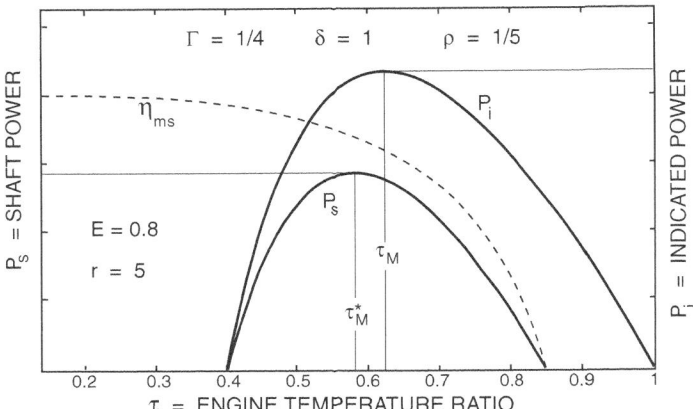

Figure 11.11 Plots of indicated power P_i and shaft power P_s as functions of temperature ratio τ for an ideal Stirling engine with limited heat transfer and internal thermal leakage.

It follows from this observation that the shaft power $P_s = P_i \, \eta_{ms}$ will always be related to indicated power P_i as shown in Figure 11.11. In particular, the *maximum shaft power point* τ_M^* can never exceed the maximum indicated power point τ_M. Thus, in general,

$$\tau_M^* \leq \tau_M \text{ with equality holding if and only if } 1/r \geq \tau_M.$$

Thus, the effect of mechanical losses on ideal Stirling engines is to move the maximum shaft power engine temperature ratio toward the reservoir temperature ratio. This is counter to the effect of a heat leak through the engine on the indicated power that was deduced above. The effects of these two losses on the location of the maximum shaft power point are thus seen to be always qualitatively opposite. Other specific examples can be found in Senft (1997).

UNIVERSAL POWER MAXIMA

The power maxima of Stirling engines with the limited heat transfer and heat leakage described above are universal in the sense to be described now. Define now an *equitable* class of engines to be the set of all engines having common values of a, T_H, Γ, δ, ρ, and r, and all having mechanisms of effectiveness E or less. Now, indicated power P is given by (11.3). Within an equitable class, this P is a function of the two variables τ and ξ, that is, $P = P(\tau, \xi)$. The variables τ and ξ are constrained by the given value of ρ as illustrated in Figure 11.8. Further, the function P_i given in (11.12) is maximal for the class, that is, $P(\tau, \xi) \leq P_i(\tau)$ for all permissible τ and ξ. For any engine in the class running at any of these τ and ξ values, its mechanical efficiency cannot exceed $\eta_{ms}(E, \tau, r)$ by Theorem (3.6). Therefore, the maximum of $P_i \eta_{ms} = P_s$ with respect to τ, which is, as shown above, the maximum brake power of an ideal Stirling engine, is the least upper bound on the brake power attainable by any engine within the equitable class.

POWER RELATIVE TO EFFICIENCY

From an external point of view, the *indicated thermal efficiency* of an engine with heat transfer (11.1) is given by

$$\eta_t = \frac{P_i}{\dot{Q}_H} = \frac{P_i}{a(T_H - T_U)} = \frac{P_i}{a T_H(1 - \xi)} \ .$$

Using (11.11) and (11.12) in this formula gives for any endoreversible engine

$$\eta_t = \frac{\delta(\tau - \Gamma)(1 - \tau) - \rho(1 + \delta\Gamma)(1 - \tau)^2}{\delta(\tau - \Gamma) - \rho(1 - \tau)^2}$$

The external *brake thermal efficiency* of an ideal Stirling engine with limited heat transfer and internal heat leakage is then the product $\eta_t \eta_{ms}$. A plot of shaft power versus brake thermal efficiency for a particular Stirling engine is shown in Figure 11.12. The plot shows that peak brake power and peak efficiency occur at distinct engine operating points.

It should be noted that the loop-shaped plot arises here primarily due to thermal considerations. The plot of indicated power versus

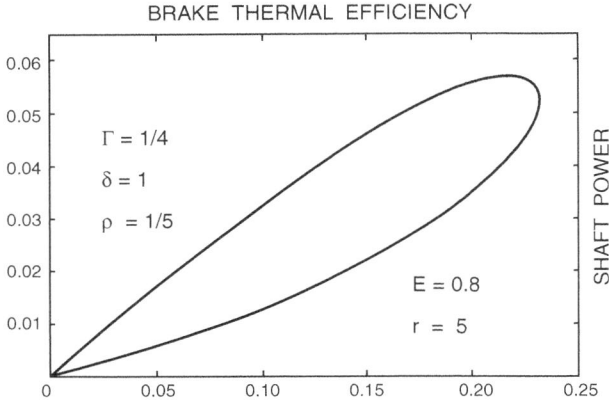

Figure 11.12 Plot of shaft power versus brake thermal efficiency for a Stirling cycle engine with limited heat transfer and internal heat leakage.

indicated thermal efficiency for the same example engine is a larger but similarly shaped loop. Indeed, the characteristics of η_{ms} noted above and illustrated in Figure 11.10 imply that the shapes must be similar. This looplike shape for indicated power versus indicated efficiency also is characteristic of the case of external heat leakage (Gordon & Huleihil, 1992).

APPENDIX A

GENERAL THEORY OF
MACHINES, EFFECTIVENESS,
AND EFFICIENCY

In this section an analytical basis for machines is developed which may be applied as a general mathematical model of the mechanical section of an engine. A universal definition of mechanism effectiveness is given. A formula is derived expressing output work in terms of input work and effectiveness.

KINEMATIC MACHINES

The simplest machine is one having two actuators by which it interacts with external agents; Figure A.1 is a schematic representation. Each actuator is a device capable of linear or angular motion, but not both, and of accepting or applying a force or torque, respectively. One dimensional coordinate systems x and y are assigned to the range of motion of each actuator.

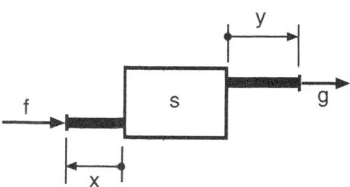

Figure A.1 Schematic representation of a two-actuator machine.

STATE PARAMETER

In kinematic machines, the connection between the actuators is most conveniently represented by a single parameter, usually s, which also serves to specify the position of all the elements of the machine. Accordingly, it is referred to as the *state parameter* of the machine. Often the state parameter can be taken to be one of the actuator positions. For example, s may be taken to be the crank angle in the analysis of a two-stroke internal combustion engine. For a four-stroke engine, the camshaft angle would be appropriate. In a typical scissors jack, the cumulative screw angle, the linear postion of the nut block, or even the angle of one of the scissor arms would make an acceptable state parameter since each of these would identify the positions of all the parts of the jack.

In particular, the actuator positions of a machine are functions of the state parameter:

$$x = x(s) \text{ and } y = y(s). \tag{A.1}$$

The state parameter is assumed to range over a closed bounded interval, and the actuator position functions (A.1) are assumed continuous in s with piecewise continuous derivatives. The operation of the machine in time is represented by a functional relationship between the state parameter s and time t :

$$s = \sigma(t).$$

The function σ is assumed twice differentiable on the associated closed bounded time interval.

ACTUATOR FORCES

Forces applied to or by a linear actuator are assumed to be coaxial with the line of actuator motion. Side loads on an actuator are not considered here. Any side loading on machine members, such as, for example, side loading on a piston due to the angular position of its connecting rod, can be modeled within the confines of the machine. Thus, the piston in, say, a four-stroke engine is a linear actuator that accepts or applies forces from or to the working gas, which are in fact collinear with the motion of the piston. The side loading on the piston is inside the mechanism section of the engine and results in internal friction loss.

Likewise, torques applied to a rotary actuator, such as an engine shaft, are taken as concentric with the axis of rotation. Any other loading, such as tension from a belt drive on the shaft, can be treated separately from the engine analysis or can be modeled as an internal load on the shaft bearings. These coaxiality assumptions simplify the description of the machine and do not unduly limit the range of application, especially for the present goal of modeling the mechanism sections of reciprocating engines and pumps.

For uniformity, energy exchanges will be expressed relative to the machine. This means that once the actuator coordinate scales have been chosen or set, *the positive external force or torque direction is taken to be the direction of increasing actuator coordinate*. This convention allows a uniform identification of the direction of workflow in time. Let f represent the force or torque acting on the x-actuator. If $f \cdot x'(s) > 0$ and $\dot{s} = \sigma'(t)$ is positive, then f is doing work on the machine in state s at time t. If, on the other hand, $f \cdot x'(s) < 0$ and $\dot{s} > 0$, then the machine is doing work against the external resistance f.

If $\dot{s} = \sigma'(t)$ is negative, then the work directions are just the opposite. Although the assumption that $\dot{s} > 0$ is all that is needed for most analyses, allowing the option that $\dot{s}$ might be negative conveniently permits one to compare the operation of reversible machines in both directions.

FORCE RELATION

If f and g are the forces or torques acting on the x- and y-actuators, respectively, of a machine at state s at a given instant, the relation between them has the general form

$$F(s, \dot{s}, \ddot{s}, f, g) = 0. \qquad (A.2)$$

This relation can be obtained in principle through a static-type force analysis with inertia forces and moments included, as well as Coulomb or viscous friction between the various machine members. It will be assumed for the machines treated here that F is continuous in its five variables.

INTERNAL ENERGY

At each state during a particular regime of operation, a machine will contain some measure of mechanical energy. Some is due, for example, to the compression of internal springs or the elevation of a machine member. The rest is due to moving parts with non-neglible mass carrying kinetic energy. The sum total of these energies is referred to here as the instantaneous *internal mechanical energy* U of the machine. It is a function of the internal positional state s and its time rate of change $\dot{s}$ since these two variables determine the velocities of all the masses within the machine:

$$U = U(s, \dot{s}). \tag{A.3}$$

In general, as a machine moves in time, its internal energy fluctuates. As its parts accelerate or decelerate, or as springs are compressed or extended, the machine accepts, stores, or releases energy in its internal parts. Usually, the potential energy component is small in engines, but the kinetic energy content can be appreciable in high-speed machines. In low-speed, highly loaded devices, such as steam engines or jacks, neither component is large, and there it can be safely assumed for most purposes that U is constant.

The primary function of a machine is to transmit work from one of its actuators to the other. Internal energy storage is usually a secondary consideration. In machine applications where appreciable energy must be cyclically stored and retrieved, separate devices, outside of the machine under consideration, are provided. In heat engines and compressors, a flywheel is the primary kinetic energy storage device, and external or buffer pressure serves as a potential energy reservoir. In typical engines, the internal energy content of the mechanism section is small compared to the energy stored in the flywheel (which in some designs may be housed in the crankcase), and the mechanical energy fluctuation of the mechanism proper can usually be ignored.

The elements described thus far suffice to characterize a machine from an external point of view. The collection of actuator functions (A.1), a force relation (A.2), and internal mechanical energy function (A.3) taken together describe a machine in a functional way:

$$M = [x, y, F, U].$$

This set of functions cannot be totally arbitrary, however, because machines must obey the laws of thermodynamics. The analytical form of this requirement for these functions will be derived below.

FORCE PROCESSES

A *force process* for the machine $M = [x, y, F, U]$ is here defined to be a triple of functions of time $\Psi = \langle \sigma(t), f(t), g(t) \rangle$ which satisfy (A.2):

$$F\big(\sigma(t), \sigma'(t), \sigma''(t), f(t), g(t)\big) = 0$$

for all t in a closed bounded interval $I = [a, b]$ of time over which the process takes place. Consistent with the sign conventions adopted above for actuator position and forces, the work done on an actuator is the time integral of the product of the force applied and its velocity (Symon, 1971). Thus, the work done by f on the x-actuator over the time interval I for the process Ψ is :

$$W\langle \sigma, f, x \rangle = \int_{\sigma} f \, dx = \int_{a}^{b} f(t) \, x'(\sigma(t)) \, \sigma'(t) \, dt$$

and similarly the work done on the y-actuator by g is

$$W\langle \sigma, g, y \rangle = \int_{\sigma} g \, dy = \int_{a}^{b} g(t) \, y'(\sigma(t)) \, \sigma'(t) \, dt.$$

The total work done on the machine M by the force process Ψ is the sum of these two actuator works:

$$W\langle \Psi, M \rangle = W\langle \sigma, f, x \rangle + W\langle \sigma, g, y \rangle.$$

FRICTIONAL DISSIPATION

Mechanical energy loss due to friction between the moving parts of a machine manifests itself as thermal energy, which is ultimately lost to the surroundings. Since this is the only exchange possible between thermal and mechanical energy in a basic machine the first law of thermodynamics mandates that in any operation of the machine the net work done on its actuators, W, cannot be less than the change in its internal energy, ΔU:

$$W \geq \Delta U. \tag{A.4}$$

The difference between the two sides of (A.4) is the energy lost to friction in the process. This is always in the context of a process which satisfies the force relation (A.2) for the machine.

   The instantaneous form of (A.4) is now derived; this is helpful for understanding the most basic characteristics of a machine because it is process independent. Let particular values $s_o, \dot{s}_o, \ddot{s}_o, f_o$ and g_o be given which satisfy (A.2). Take any function $\sigma(t)$ defined on a time interval such that at some particular instant t_o

$$\sigma(t_o) = s_o$$
$$\sigma'(t_o) = \dot{s}_o$$
$$\sigma''(t_o) = \ddot{s}_o$$

Choose also a function $f(t)$ so that $f(t_o) = f_o$ and then let $g = g(t)$ be the solution of

$$F\big(\sigma(t), \sigma'(t), \sigma''(t), f(t), g\big) = 0$$

on a subinterval containing t_o. It is assumed that F is, and that $\sigma(t)$ and $f(t)$ are chosen, such that this is possible. For example, the hypothesis

$$\left. \frac{\partial F}{\partial g} \right|_{s_o, \dot{s}_o, \ddot{s}_o, f_o} \neq 0$$

of the Implicit Function Theorem (Buck, 1965) would be sufficient to solve (A.2), namely $F(s, \dot{s}, \ddot{s}, f, g) = 0$, for $g = g(s, \dot{s}, \ddot{s}, f)$ in a neighborhood of $(s_o, \dot{s}_o, \ddot{s}_o, f_o)$.

Letting $\Delta t > 0$ be a postive increment of time, Condition (A.4) gives

$$\int_{t_o}^{t_o + \Delta t} \left[f(t) \, x'(\sigma(t)) + g(t) \, y'(\sigma(t)) \right] \sigma'(t) \, dt \geq$$

$$U\big(\sigma(t_o + \Delta t), \sigma'(t_o + \Delta t)\big) - U\big(\sigma(t_o), \sigma'(t_o)\big).$$

Assuming the required conditions are satisfied, applying the Mean Value Theorem to both sides yields

$$\left[f(t_1) \, x'(\sigma(t_1)) + g(t_1) \, y'(\sigma(t_1)) \right] \sigma'(t_1) \, \Delta t \geq \left[\frac{d}{dt} U(\sigma(t), \sigma'(t)) \right]_{t_2} \Delta t$$

where $t_o \leq t_1, t_2 \leq t_o + \Delta t$. Dividing by Δt, which was assumed positive, removes it from the inequality but preserves the sense of the inequality. The right-hand side is

$$\frac{\partial U}{\partial s}\big(\sigma(t_2), \sigma'(t_2)\big) \, \sigma'(t_2) + \frac{\partial U}{\partial \dot{s}}\big(\sigma(t_2), \sigma'(t_2)\big) \sigma''(t_2).$$

Letting $\Delta t \to 0$ forces $t_1, t_2 \to t_o$ which gives the inequality

$$\left[f(t_o) \, x'(\sigma(t_o)) + g(t_o) \, y'(\sigma(t_o)) \right] \sigma'(t_o) \geq$$

$$\frac{\partial U}{\partial s}\big(\sigma(t_o), \sigma'(t_o)\big) \, \sigma'(t_o) + \frac{\partial U}{\partial \dot{s}}\big(\sigma(t_o), \sigma'(t_o)\big) \sigma''(t_o)$$

and using the original state values at t_o yields

$$f_o \, x'(s_o) \, \dot{s}_o + g_o \, y'(s_o) \, \dot{s}_o \geq \frac{\partial U}{\partial s}(s_o, \dot{s}_o) \, \dot{s}_o + \frac{\partial U}{\partial \dot{s}}(s_o, \dot{s}_o) \, \ddot{s}_o.$$

Dividing by $\dot{s}_o$ gives the law governing frictional power dissipation:

> For any values of $s, \dot{s}, \ddot{s}, f,$ and g that satisfy the force relation $F(s, \dot{s}, \ddot{s}, f, g) = 0$ of a machine $M = [x, y, F, U]$,
>
> $$\text{if } \dot{s} > 0 \text{ then } \quad f \, x'(s) + g \, y'(s) \geq \frac{\partial U}{\partial s} + \frac{\partial U}{\partial \dot{s}} \frac{\ddot{s}}{\dot{s}}. \qquad (A.5)$$
>
> If $\dot{s} < 0$ then the inequality has the opposite sense.

GRAPHICAL REPRESENTATION

Inequality (A.5) has a straightforward graphical interpretation. The curve shown in Figure A.2 is the imagined graph of actuator force or torque g versus f based upon a force relation $F(s, \dot{s}, \ddot{s}, f, g) = 0$ for fixed or instantaneous values of s, $\dot{s}$, and $\ddot{s}$.

The right-hand side of (A.5) is denoted by D in the figure:

$$D = \frac{\partial U}{\partial s} + \frac{\partial U}{\partial \dot{s}} \frac{\ddot{s}}{\dot{s}}. \tag{A.6}$$

Physically, this is the rate of internal energy change with respect to state. The line of slope $m_v = -x'/y'$ with g-intercept D/y' represents the working of the machine with no friction loss. Inequality (A.5) requires the curve $F(s, \dot{s}, \ddot{s}, f, g) = 0$ to be in one of the half planes determined by this line. For the case when $\dot{s}$ and $y'(s)$ are positive, it is the upper halfplane, as shown in Fig. A.2.

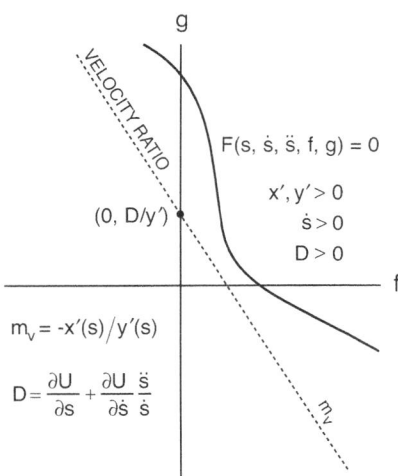

Figure A.2 Graph of a general force relation for instantaneous values of s, $\dot{s}$, and $\ddot{s}$.

REVERSED OPERATION

To clarify the concepts introduced above, the implication of (A.5) when the same machine is operated in the reversed direction will now be illustrated. The exact same motion in time is taken but opposite in direction. If the one motion is $s = \sigma(t)$ over a time interval $[\alpha, \beta]$, then its direct reversal can be described by $s_R(t) = \sigma(-t)$ over the time interval $[-\beta, -\alpha]$. In this parameterization of the two directions of operation, at time $-t$ in the reversed motion the machine state is the same as it is at time t in the forward motion:

$$s(t) = s_R(-t).$$

Hence,

$$\dot{s}(t) = -\dot{s}_R(-t)$$

and

$$\ddot{s}(t) = \ddot{s}_R(-t).$$

That is, at corresponding times t and $-t$, in the forward and reversed motions, the states are the same, state velocities are opposite, and state accelerations are the same.

Now the internal mechanical energy (A.3) is a function of state and state velocity. In the kind of machines usually encountered, the function U does not depend upon the sign of the state velocity, that is,

$$U(v, w) = U(v, -w).$$

Therefore,

$$\frac{\partial U}{\partial s}(v, w) = \frac{\partial U}{\partial s}(v, -w)$$

and

$$\frac{\partial U}{\partial \dot{s}}(v, w) = -\frac{\partial U}{\partial \dot{s}}(v, -w).$$

Consider now the function $D = D(s, \dot{s}, \ddot{s})$ defined by (A.6) and calculate as follows:

$$
\begin{aligned}
D(v, w, z) &= \frac{\partial U}{\partial s}(v, w) + \frac{\partial U}{\partial \dot{s}}(v, w)\frac{z}{w} \\
&= \frac{\partial U}{\partial s}(v, -w) - \frac{\partial U}{\partial \dot{s}}(v, -w)\frac{z}{w} \\
&= \frac{\partial U}{\partial s}(v, -w) + \frac{\partial U}{\partial \dot{s}}(v, -w)\frac{z}{-w} = D(v, -w, z).
\end{aligned}
$$

This means that at corresponding times in the forward and reversed motion of the machine, the right-hand side of (A.5) is the same for both directions. However, the sense of the inequality is reversed because $\dot{s}(t) = -\dot{s}_R(-t)$. Therefore, in the graphical representation of (A.5) the velocity line is the same for both forward and reversed motion, but the force relations will be in opposite half-planes. Figure A.3 shows a hypothetical force relation having the same velocity ratio line as in Figure A.2 but with $\dot{s} < 0$. Note that the shape and characteristics of the force relation will in general be different for forward and reversed operation.

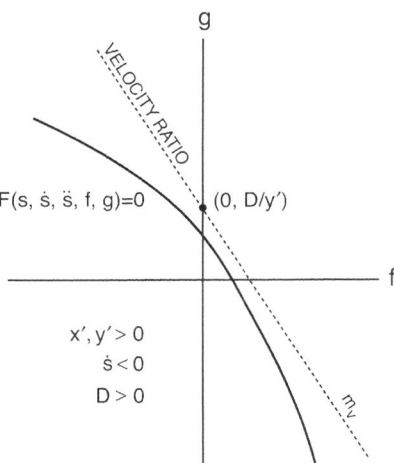

Figure A.3 The force relation for the same hypothetical machine of Figure A.2 but with reversed motion.

MECHANISM EFFECTIVENESS

If $f, g, s, \dot{s}$, and $\ddot{s}$ are values satisfying the force relation (A.2) for machine $M = [x, y, F, U]$, set

$$A = f \, x'(s) \, \dot{s}$$
$$B = g \, y'(s) \, \dot{s}$$
$$C = D \dot{s}$$

where D is given by (A.6). The instantaneous efficiency or *mechanism effectiveness* ε of M at this point is defined as

$$\varepsilon = \frac{A^- + B^- + C^+}{A^+ + B^+ + C^-}. \tag{A.7}$$

The positive and negative part functions used here are as in standard mathematical practice (Rudin, 1964):

$$z^+ = \max\{0, z\} \quad \text{and} \quad z^- = \max\{0, -z\}.$$

Note that both parts are non-negative by definition and that the following is an identity:

$$z = z^+ - z^- \tag{A.8}$$

Effectiveness ε is defined to be zero whenever $A = B = C = 0$.

The definition of effectiveness given here is a natural one. The term A equals the time rate of work being done on the x-actuator, and B is that being done on the y-actuator. The term C is the time rate of change in the internal energy of the machine. These terms may be positive or negative, of course; Definition (A.7) sorts out how these quantities should be tallied in accounting the effectiveness of the machine as a work transducer.

In the denominator of (A.7) are the quantities which the machine has at its disposal to use to do work. If $A > 0$, then A^+ in the denominator counts it as power being delivered to the machine through the x-actuator. This the machine may use to do external work through the other actuator, or to increase its internal energy store, or some of both. Similarly, B^+ adds to the denominator any power being delivered to the machine through the y-actuator. C^- counts in any power the machine may be releasing from its stored internal energy.

In the numerator of ε are the quantities that "result" from those in the denominator. If $A < 0$, then A^- appears in the numerator as the rate at which the machine is doing external work through the x-actuator. This is likewise for B^- and the y-actuator. C^+ is the rate at which the internal energy of the machine is increasing.

Thus, ε is a ratio measure of how well a machine utilizes power. In the denominator it puts whatever it receives through an actuator or what it takes from its internal reservoir. It places in the numerator whatever it transfers out through another actuator or whatever it puts into internal storage.

If inequality (A.5) is multiplied through by $\dot{s}$, it takes the simple form

$$A + B - C \geq 0$$

for both positive and negative $\dot{s}$. This implies ε is always defined. Further, using (A.8) on this gives

$$A^+ + B^+ + C^- \geq A^- + B^- + C^+$$

which shows that $\varepsilon \leq 1$. Therefore, $0 \leq \varepsilon \leq 1$ is always true.

If $\dot{s} > 0$ then effectiveness can be expressed in a simpler form.

$$\text{If } \dot{s} > 0, \quad \text{then} \quad \varepsilon = \frac{[f \cdot x']^- + [g \cdot y']^- + D^+}{[f \cdot x']^+ + [g \cdot y']^+ + D^-}. \tag{A.9}$$

If $\dot{s} < 0$ then ε turns out to be the reciprocal of the quotient in (A.9). Throughout most applications, $\dot{s}$ will be of one sign, as, for example, in studies of the steady operation of heat engines. In such situations where the sign of $\dot{s}$ will not change, the state parameter can be selected so that $\dot{s} > 0$, and then (A.9) can be used for ε.

The value and usefulness of the effectiveness function defined here lie first of all in its intuitive content. It basically is a single number whose magnitude reflects how much of the energy that a machine takes is put to good use. This is an easy thing for the mind to grasp.

Beyond this, it has analytic value also inasmuch as from this single number can be extracted essentially all of the relation between the actuator forces, as will be shown in the following section.

CONTENT OF THE EFFECTIVENESS FUNCTION

Suppose particular values of $s, \dot{s}, \ddot{s}, x', y', f, D$, and ε are given for a particular machine. It will be shown that if $\varepsilon \neq 0$ then the value of the force or torque g acting on the y-actuator is uniquely determined. By Definition (A.7),

$$\left(A^+ + B^+ + C^-\right)\varepsilon = A^- + B^- + C^+ \qquad (A.10)$$

is certainly true when $\varepsilon \neq 0$. But $\varepsilon = 0$ if and only if $A, B \geq 0$ and $C \leq 0$, so (A.10) is true in all cases. The goal now is to solve (A.10) for $B = g\, y'(s)\, \dot{s}$, which will yield g since $\dot{s}$ and $y'(s)$ are given.

Letting $a = A^+ + C^-$ and $b = A^- + C^+$ Equation (A.10) takes the form

$$\varepsilon\left(a + B^+\right) = b + B^-$$

which is equivalent to

$$\varepsilon\, a - b = B^- - \varepsilon\, B^+. \qquad (A.11)$$

Equation (A.11) can be solved when $\varepsilon \neq 0$ by considering the various cases for the sign of the left-hand side.

Case 1: $\varepsilon\, a - b = 0$ and $\varepsilon \neq 0$. In this case, (A.11) reduces to $\varepsilon\, B^+ = B^-$, which since $\varepsilon \neq 0$ has

$$B = 0$$

as its only solution.

Case 2: $\varepsilon\, a - b > 0$ and $\varepsilon \neq 0$. In this case, (A.11) implies $\varepsilon\, B^+ < B^-$, which only can be true if $B < 0$ since $\varepsilon \neq 0$. This reduces (A.11) to the equation $\varepsilon\, a - b = B^- = -B$ so

$$B = b - \varepsilon\, a.$$

Case 3: $\varepsilon\,a - b < 0$ and $\varepsilon \neq 0$. Here (A.11) first implies $\varepsilon\,B^+ > B^-$ and this implies $B > 0$. This reduces (A.11) to $\varepsilon\,a - b = -\varepsilon\,B$, and because $\varepsilon \neq 0$ this can be solved for B :

$$B = \frac{b - \varepsilon\,a}{\varepsilon}.$$

These cases show (A.11) uniquely determines B when $\varepsilon \neq 0$. This is not so when $\varepsilon = 0$, for then (A.11) requires only that

$$-b = B^-.$$

But when $\varepsilon = 0$, the definition implies $b = 0$ also. Thus, any $B \geq 0$ would satisfy (A.11), so no unique value of B is determined in this case.

Combining the results of the cases above yields the following formula for B:

$$B = \frac{[\varepsilon\,a - b]^-}{\varepsilon} - [\varepsilon\,a - b]^+ \quad \text{when } \varepsilon \neq 0. \tag{A.12}$$

The next section will show the utility of Equation (A.12) for expressing the work done by one actuator of a machine for a given work process applied to the other.

ACTUATOR WORK

Consider now a machine $M = [x, y, F, U]$ and a force process $\Psi = \langle \sigma(t), f(t), g(t) \rangle$ given for the machine over a closed bounded interval I of time. In this setting, the parameters A, B, and C defined above are functions of t over the interval $I = [a, b]$:

$$A = f(t)\,x'\big(\sigma(t)\big)\,\sigma'(t)$$
$$B = g(t)\,y'\big(\sigma(t)\big)\,\sigma'(t)$$
$$C = \frac{\partial U}{\partial s}\big(\sigma(t), \sigma'(t)\big)\,\sigma'(t) + \frac{\partial U}{\partial s}\big(\sigma(t), \sigma'(t)\big)\,\sigma''(t).$$

We assume M and Ψ are such that A, B, and C are piecewise continuous and of finite oscillation on I. The latter condition means that the functions change from a condition of being positive, zero, or negative to another of these conditions only a finite number of times on the interval. These conditions allow I to be partitioned into endpoint-overlapping closed subintervals such that

- A, B, and C are continuous on the interior of each subinterval and are continuously extendable to the endpoints, and
- throughout the interior of each subinterval,
  exactly one of $A > 0$, $A = 0$, or $A < 0$ holds,
  and exactly one of $B > 0$, $B = 0$, or $B < 0$ holds, and
  exactly one of $C > 0$, $C = 0$, or $C < 0$ holds.

Subintervals on the interior of which A, B, $-C \geq 0$ will be called Type 1. Type 2 intervals are the others, that is, those on the interior of which at least one of A, B, or $-C$ is negative. In this context, ε is a function of t. On the interior of Type 2 intervals, $\varepsilon \neq 0$, and on the interior of Type 1 intervals, $\varepsilon = 0$. It follows that any other point where $\varepsilon = 0$ must be an endpoint of one of the subintervals of the partition. Let J be the union of all Type 1 intervals and K be the union of all Type 2 intervals. Then $I = J \cup K$ with overlap at subinterval endpoints only.

Let $W_o \langle \sigma, g, y \rangle$ denote the work output of the y-actuator. Then

$$W_o \langle \sigma, g, y \rangle = -W \langle \sigma, g, y \rangle = -\int_I B\, dt = -\int_J B\, dt - \int_K B\, dt. \quad \text{(A.13)}$$

The first integral following the rightmost equal sign in (A.13) will be denoted by W_n and called *null work*:

$$W_n \langle \sigma, g, y \rangle = \int_J B\, dt.$$

Null work is work done on the y-actuator whenever the effectiveness of the mechanism is zero.

Using (A.8) on the last integral in (A.13), gives

$$W_o \langle \sigma, g, y \rangle = \int_K B^-\, dt - \int_K B^+\, dt - W_n. \quad \text{(A.14)}$$

By (A.12), $B^- = [\varepsilon\, a - b]^+$ on the interior of K, and therefore

$$\int_K B^-\, dt = \int_K [\varepsilon\, a - b]^+\, dt.$$

On the interior of J, $\varepsilon = 0$, which implies also that $b = 0$, as already noted. Therefore, $\varepsilon a - b = 0$ on the interior of J, and so

$$\int_J [\varepsilon a - b]^+ \, dt = 0.$$

Thus

$$\int_K B^- \, dt = \int_I [\varepsilon a - b]^+ \, dt \tag{A.15}$$

Also by (A.12), $B^+ = [\varepsilon a - b]^- / \varepsilon$ on the interior of K. By the assumed piecewise continuity of B, we may then write

$$\int_K B^+ \, dt = \int_K \frac{[\varepsilon a - b]^-}{\varepsilon} \, dt$$

however the integrand on the right-hand side may be defined at interval endpoints (where ε may be zero). For the sake of definiteness, we may define a special division operator as

$$1/\!/z = \frac{1}{z} \quad \text{if} \quad z \neq 0$$
$$= 0 \quad \text{if} \quad z = 0.$$

Then

$$\int_K B^+ \, dt = \int_K (1/\!/\varepsilon) [\varepsilon a - b]^- \, dt.$$

Also, $\int_J [\varepsilon a - b]^- \, dt = 0$ as before. Therefore

$$\int_K B^+ \, dt = \int_I (1/\!/\varepsilon) [\varepsilon a - b]^- \, dt. \tag{A.16}$$

Assembling (A.14)-(A.16) yields the following formula:

$$W_o \langle \sigma, g, y \rangle = \int_I [\varepsilon a - b]^+ \, dt - \int_I (1/\!/\varepsilon) [\varepsilon a - b]^- \, dt - W_n. \tag{A.17}$$

This formula gives the harder way of computing output of the y-actuator when a complete force process is known, for the quantity in question is simply $-\int_\sigma g \, dy = -\int_I B \, dt$, which can be computed directly if the y-actuator force g is known. But Formula (A.17) becomes useful when internal energy is constant and the variation of machine state in time is known or given, for then the y-actuator output can be expressed and easily bounded in terms of what is applied to the x-actuator and the effectiveness of the machine.

CONSTANT INTERNAL ENERGY

In many instances, the variation of energy contained within the mechanism is small and can often be neglected for the purposes at hand. An example would be an engine having lightweight reciprocating members operating at a constant speed. The main mechanical energy store for such an engine would be a flywheel, which can be regarded and modeled as external to the engine mechanism proper. Another example would be a slowly moving engine or pump, where the kinetic energy content is small and may be regarded as zero, a constant.

Whenever internal mechanical energy can be regarded as constant, $D = 0$. Therefore, $C = D\dot{s} = 0$ and the following reductions can be made:

$$[\varepsilon\, a - b]^+ = \left[\varepsilon A^+ - A^-\right]^+ = \left[\varepsilon A^+\right]^+ = \varepsilon\, A^+$$

and

$$[\varepsilon\, a - b]^- = \left[\varepsilon A^+ - A^-\right]^- = A^-.$$

Therefore (A.17) reduces to

$$W_o\langle \sigma, g, y\rangle = \int_I \varepsilon\, A^+\, dt - \int_I (1/\!/\varepsilon)\, A^-\, dt - W_n \qquad \text{(A.18)}$$

whenever internal energy is constant.

If it is known that the effectiveness ε is bounded above by a constant E over the entire force process, then (A.18) shows

$$W_o\langle \sigma, g, y\rangle \leq \mathrm{E} \int_I A^+\, dt - \frac{1}{\mathrm{E}} \int_I A^-\, dt - W_n.$$

The two integrals on the right-hand side represent the two kinds of work that the x-actuator experiences from the f-component of the force process. The first is *efficacious work* done on the x-actuator:

$$W_+\langle \sigma, f, x\rangle = \int_\sigma [f\, dx]^+ = \int_a^b [f(t)x'(\sigma(t))\sigma'(t)dt]^+.$$

The second integral in the inequality above is the *forced work* done by the x-actuator moving against the resisting force f :

$$W_-\langle \sigma, f, x \rangle = \int_\sigma [f \, dx]^- = \int_a^b [f(t)x'(\sigma(t))\sigma'(t)dt]^- .$$

This is summarized by the following:

If E is a constant such that $E \geq \varepsilon$ over the force process $\langle \sigma, f, g \rangle$ and if the internal energy of the machine is constant over the force process, then

$$W_o\langle \sigma, g, y \rangle \leq E \, W_+\langle \sigma, f, x \rangle - \frac{1}{E} W_-\langle \sigma, f, x \rangle - W_n. \quad (A.19)$$

Since

$$W_+\langle \sigma, f, x \rangle - W_-\langle \sigma, f, x \rangle = W\langle \sigma, f, x \rangle$$

the conclusion of the above result can also be written as

$$W_o\langle \sigma, g, y \rangle \leq E \, W\langle \sigma, f, x \rangle - \left(\frac{1}{E} - E \right) W_-\langle \sigma, f, x \rangle - W_n. \quad (A.20)$$

For situations in which the effectiveness of the machine has a constant value $E \neq 0$ throughout the process, (A.18) yields the equality

$$W_o = E \, W - \left(\frac{1}{E} - E \right) W_- . \quad (A.21)$$

AN ULTRA LOW TEMPERATURE DIFFERENTIAL STIRLING ENGINE

Although the analysis presented in Chapter 7 is highly idealized, it is quite appropriate for providing some insight into the geometrical requirements of the ultra low temperature differential Stirling engine illustrated in Figures B.1–B.3. Nicknamed the P-19, this engine has proven itself capable of operating down to a temperature difference of just 0.5 °C (less than 1 °F) between its warm and cool sides. The P-19 was the first to run from heat absorbed while resting on the palm of a human hand. The P-19 was first publicly demonstrated at the 25th Intersociety Energy Conversion Engineering Conference held in Reno, Nevada, in August 1990

BACKGROUND

A low temperature differential (LTD) Stirling engine may be characterized as one that operates more or less optimally with a temperature difference of less than 100 °C between its hot and cold end. Ivo Kolin was the first to design and build such an engine. At the Inter-University Center in Dubrovnik in 1983 he demonstrated the first of his engines operating with hot water as the heat source and cold water as the heat sink (Kolin, 1983). The engine continued to run until the temperature difference between the source and sink dropped to 15 °C.

Kolin's first engine inspired a number of research projects over the next decade to further develop LTD Stirling engines (Senft, 1996).

Figure B.1 A photograph of the ultra low temperature differential engine P-19 showing the power cylinder unit mounted on the top of the large flat displacer chamber.

LTD engines were designed and built during this period to study Ringbom engine dynamics, regenerator effectiveness, passive solar operation, and mechanical efficiency. These engines excelled as research tools and instructional aids to demonstrate Stirling and general heat engine principles. In fact, the basic ideas for the general theory of mechanical efficiency were inspired from observations of the behavior of these engines, with the first paper on the subject appearing in 1985 (Senft, 1985).

In all honesty, the P-19 engine project was primarily motivated by the desire to make an engine that could run on the smallest temperature

Figure B.2   The P-19 engine is shown here operating on the slight cooling effect provided by a cold drink placed on its top plate.

difference then known. Of course, this is a dubious goal from a strictly engineering point of view because efficiency and specific power both suffer as the temperature differential becomes smaller. Nevertheless, it was an intriguing challenge that could be adequately justified on educational and scientific grounds. It was envisioned as an excellent teaching device and would concretely illustrate the radically low compression ratio that the then-new mechanical efficiency theory, described in the pages of this book, pointed to as necessary for ultra LTD operation.

Details of the construction and performance of the P-19 have been published elsewhere (Senft, 1995, 1996). The design strategy has also been described in the literature (Senft, 1991c, 1992). A key aspect of the design was choosing the compression ratio for the engine, but because several other parameters are unknown for this type of engine, it is only possible to analytically bracket the compression ratio range. The P-19 engine was made so that the stroke of the piston could be changed and the best compression ratio chosen experimentally.

COMPRESSION RATIO LIMITS

Recall from Chapter 4 that if the square of mechanism effectiveness is less than the temperature ratio, then there is a limit to how high a compression ratio can be used. For the usual flame-heated Stirling engine, τ is around 0.35 or 0.4: it would take a very poor mechanism indeed to satisfy $E^2 < \tau$, and therefore there is no theoretical restriction on compression ratio. But for a very LTD engine, τ is quite high. Table B.1 here shows the temperature ratios for some low ΔT values based upon $T_H = 300$ K. This is an appropriate hot-side temperature for an engine using a human hand or similar low-grade heat source in normal indoor ambient conditions. It is clear from the table that with even with an outstanding mechanism effectiveness of, say, $E = 0.95$, E^2 will be significantly less than all the table values of τ. Therefore, the usable compression ratios are indeed restricted by theory for very LTD engines.

The analysis of Chapter 7 can be used to give a good idea of just how restricted compression ratio is at these low ΔT values. In fact, the analysis there is quite appropriate for an ultra LTD engine. Slow-speed operation is intrinsic to LTD engines, especially for one running at just a degree or two differential. Engine speed is proportional to the rate of heat transfer into and out of the engine which in turn is proportional to the difference in temperature between the engine gas and the engine hot and cold surfaces. If the difference between these surfaces is small, the rate of heat transfer and so engine speed will be low.

B.1 *Table of temperature ratios for LTD engines*

ΔT °C	$\tau = T_C / T_H$ $T_H = 300\,K$
9	0.970
6	.980
3	.990
2	.993
1	.997
0.5	.998

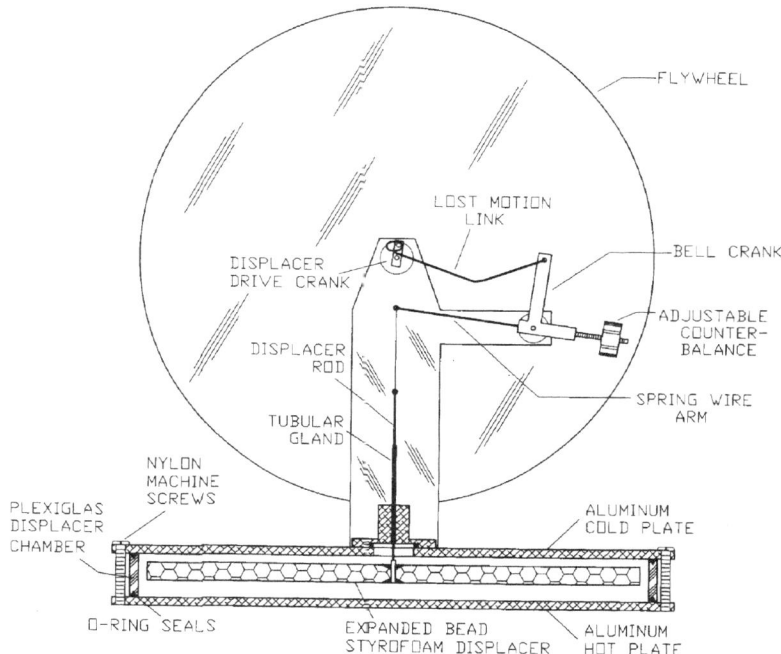

Figure B.3 This scale drawing of the P-19 shows the displacer drive system. The weight of the displacer is counterbalanced by a lever pivoted on ball bearings. The displacer is driven through a bell crank arm attached to this lever. Dwell periods at the ends of the displacer stroke are produced by the loop in the connecting link for a lost motion effect.

To give the engine some aid in effecting its heat transfers, a displacer drive with dwell periods at the extremes of its stroke was built into the P-19, as shown in Figure B.3. This intermittent drive gives the engine cycle a definite resemblance to the four-step ideal cycles modeled in Chapter 7.[†] When observing the smallest temperature difference at which the engine will operate, it is turning over very slowly with very long displacer dwell periods, and therefore this resemblance is more marked as ΔT gets smaller and smaller. Hence, the results of Chapter 7 could well be expected to be a good guide in this case.

[†] This has been experimentally verified for LTD Ringbom Stirling engines which also have a long displacer dwell at lower speeds (Senft, 1993a, Chap. 8).

MEAN VOLUME SPECIFIC WORK

A rather minor alteration that one might wish to make to the analysis is to use mean engine volume rather than the maximum engine volume for specific work. This is because, if the engine is made with the piston stroke adjustable by changing the radius at which the crankpin is set and the connecting rod is fixed in length, the mean volume of the engine will essentially remain the same. Now it is true that more dead volume will be present with the shorter piston strokes, but the large displacer chamber volume makes its effect quite negligible.

In fact, the displacer swept volume is so large in the P-19 (about 225 cc) that there is little difference between the mean and total or maximum volume of the engine at the piston strokes used (1 or 2 cc). But for the sake of mathematical variety, the mean volume will be used here for specific work. The *mean volume specific shaft work* is defined as

$$\check{W}_s = \frac{W_s}{p_m V_C} = \frac{W \eta_{mc}}{p_m V_C}$$

where $V_C = (V_m + V_M)/2 = V_M - \Delta V/2$ is the mean engine volume and all other parameters are as before. Since $V_C = V_M(1 + r)/(2r)$, comparison with Equation (7.18) shows

$$\check{W}_s = \frac{2r}{1 + r} \, \hat{W}_s$$

and so the results of Chapter 7 are modified to suit the present purpose by the added factor $2r/(1 + r)$, which is very near unity for the compression ratios of interest here.

Figures B.4–B.6 show graphs of this specific shaft work $\check{W}_s$ with respect to r for $E = 0.8$ and various λ and τ values. Recall that $\lambda = 1$ corresponds to the ideal or isothermal Stirling, whereas $\lambda = 1.4$ represents the ideal Otto or adiabatic Stirling. The best estimate of λ to take for this or any other LTD engine is unknown. Also unknown is the best estimate for the average mechanism effectiveness, but $E = 0.8$ appears to be a prudent choice for the first go at analysis. All three figures are drawn to the same scale for easy comparison.

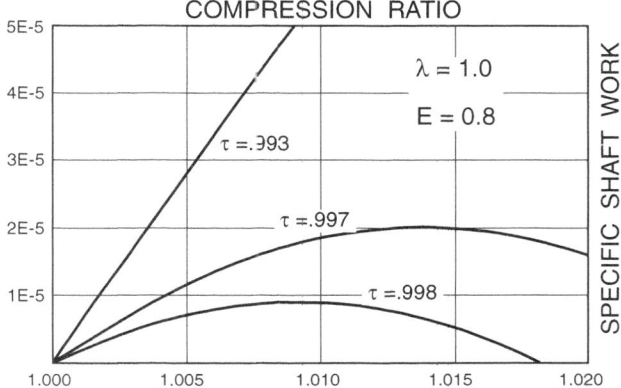

Figure B.4 Graphs of specific shaft work $\check{W}_s$ versus compression ratio r for an ideal Stirling engine with constant mechanism effectiveness E $= 0.8$ and temperature ratios τ corresponding to $\Delta T = 0.5, 1,$ and 2 °C.

Figure B.4 is the isothermal case. Curves are shown there for the three temperature ratios $\tau = 0.993, 0.997,$ and $0.998,$ which correspond respectively to temperature differentials of about 2, 1, and 0.5 °C. The isothermal case is quite optimistic even at low speeds, but it definitely provides an upper bound. For operation at $\Delta T = 0.5$ °C, the curve for $\tau = 0.998$ shows that compression ratio must be less than 1.018, and optimum would be around 1.010.

Figure B.5 corresponds to a situation intermediate between isothermal and adiabatic. In this case, a compression ratio of 1.010 would yield near optimum operation at $\Delta T = 2$ °C ($\tau = .993$). Operation would also be possible for lower ΔT, but not down to 1 °C ($\tau = .997$). For $\Delta T = 1$ °C, the optimum compression ratio is 1.005, and this would just permit operation at 0.5 °C ($\tau = .998$) as well.

Figure B.6 represents the adiabatic case. It is interesting that even here a compression ratio of 1.010 would permit operation at $\Delta T = 2$ °C, but of course not at 1 degree. A ratio of 1.005 would give better operation at 2 °, and would just barely allow operation at $\Delta T = 1$ °C, but

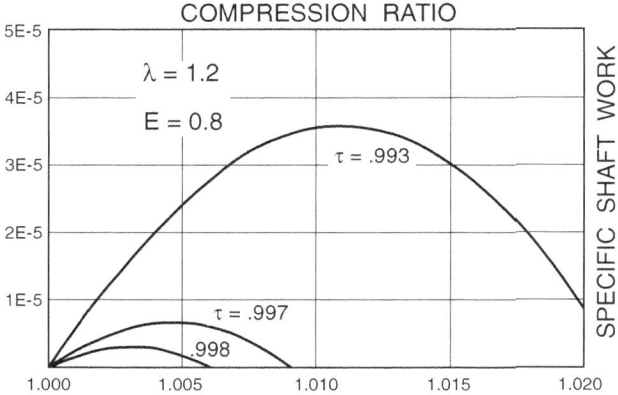

Figure B.5 Graphs as in Figure B.4 for a Crossley cycle engine with polytropic index $\lambda = 1.2$.

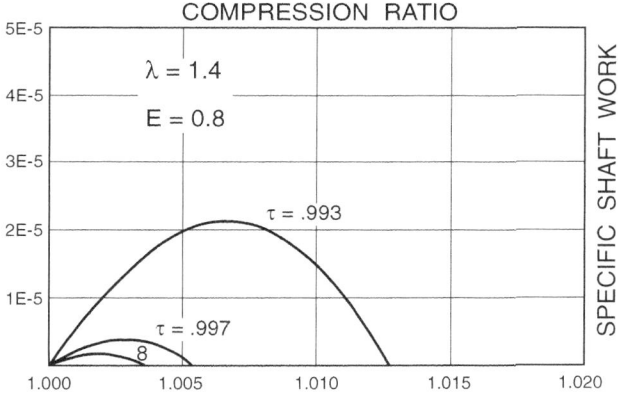

Figure B.6 Graphs as in Figure B.4 for an adiabatic Crossley cycle engine, $\lambda = 1.4$.

not at $0.5°$. These graphs indicate rather well in what range compression ratio must lie for ultra LTD operation.

ENGINE PERFORMANCE

The P-19 has been extensively operated at piston swept volumes of 2 cc and 1 cc. At 1 cc, the actual compression ratio is 1.0039 and the engine has been repeatedly observed to run with a temperature difference of only 0.5 °C between its hot and cold plates. The plate temperatures are monitored by thermocouples fitting in holes drilled into the plate edges. The accuracy of the thermometer system used is 0.05 °C. Actually, when the engine was brand new, it operated at a ΔT of 0.4 °C. After considerable use the performance has degraded only slightly and, when last tried, it still ran at $\Delta T = 0.5$ °C, but not below that at the 1 cc piston stroke. No smaller stroke has been tried.

The engine operates very well for demonstrations with the piston swept volume set at 2 cc. The compression ratio at this setting is 1.0078, and the engine begins running at a ΔT of 0.8 °C. With a temperature differential of 6 °C across its plates, the engine runs at more than 100 rpm when equipped with the regenerative displacer shown in Figure B.7; this is the temperature differential that can typically be maintained steady state when the engine is handheld in most indoor settings.

The engine will operate for hours on a few cups of hot tap water in an insulated container beneath the engine. A cup or a can containing a cold drink placed on the top plate of the engine will also run it very nicely as illustrated in Figure B.2. With the phasing readjusted, the engine will run when placed in a sunny location.

Figure B.7 The displacer of the P-19 engine fitted with regenerative elements made from polyurethane filter material.

The P-19 has also been operated on the evaporative cooling effect of water. Set up in the lab with cloth pads on the top plate kept moist by wick feed from two cups of water, the engine once ran nonstop for 16 days. During this run, the temperature difference across the plates was between 1.7 and 2.0 °C. The engine was finally stopped because it was needed by a student for a seminar demonstration. More details of the design, analysis, performance, and construction of the P-19 can be found in the references already cited.

Overall, the observed performance of the engine indicates that the intermediate case of E = 0.8 and λ = 1.2 in Figure B.5 fits the P-19 reasonably well. Of course, it must be kept in mind that these graphs are all based on the guess that mechanism effectiveness averages 80%. If the mechanism is actually better or worse, then the accompanying λ value would be higher or lower. It can be concluded at least from Figure B.6 that E > 0.8 or λ < 1.4 because the engine actually runs at a Δ T of 0.8 °C with a compression ratio of 1.0078. The same conclusion follows from the observation that the engine operates at a Δ T of 0.5 °C with a compression ratio of 1.0039. Beyond this, no serious attempts have been made to determine the values of λ and E that fit the engine best.

DERIVATION OF SCHMIDT GAMMA EQUATIONS

The following derivation supplies the equations needed for Chapter 10 and follows the nomenclature described there.

VOLUME AND PRESSURE FUNCTIONS

The instantaneous volume of gas in the hot and cold spaces of the engine are

$$V_H = \frac{V_1}{2}(1 + \cos(\omega t + \alpha))$$

and

$$V_C = V_1 - V_H + \frac{V_2}{2}(1 + \cos \omega t)$$
$$= \frac{V_1}{2}\big[1 + \kappa(1 - \cos \omega t) - \cos(\omega t + \alpha)\big].$$

The instantaneous total engine volume V is then

$$V = V_H + V_C + V_D = V_1 + \frac{V_2}{2}(1 + \cos \omega t) + V_D$$
$$= V_1\left[1 + \frac{\kappa}{2}(1 + \cos \omega t) + \chi\right].$$

In terms of the total swept volume $V_T = V_1 + V_2 = V_1(1 + \kappa)$ of the piston and displacer,

$$dV = -\frac{\omega \kappa V_T}{2(1 + \kappa)}(\sin \omega t)\, dt.$$

By the assumptions made, instantaneous pressure within the engine is

$$p = \frac{m\,R}{V_H/T_H + V_C/T_C + V_D/T_D} = \frac{m\,R\,T_C}{\tau V_H + V_C + V_1 \chi 2\tau/(1+\tau)}$$

where m is the total mass of gas captive within the engine and R is its ideal gas constant. Putting the expressions obtained above for V_H and V_C in the denominator of the last expression gives the following expression for the instantaneous pressure:

$$p =$$

$$\frac{2\,m\,R\,T_C}{V_1\left[\tau(1+\cos(\omega t + \alpha)) + 1 + \kappa(1 + \cos\omega t) - \cos(\omega t + \alpha) + 4\chi\tau/(1+\tau)\right]}$$

The factor in the brackets in the denominator can be written in the following form:

$$\left(1 + \tau + \kappa + \frac{4\chi\tau}{1+\tau}\right) + (\tau - 1)\cos(\omega t + \alpha) + \kappa\cos\omega t =$$

$$= Y + A\cos\omega t + B\sin\omega t$$

where

$$Y = 1 + \tau + \kappa + \frac{4\chi\tau}{1+\tau}, \quad A = \kappa - (1 - \tau)\cos\alpha \quad \text{and} \quad B = (1 - \tau)\sin\alpha$$

Let

$$\theta = \arccos\frac{A}{\sqrt{A^2 + B^2}} \quad \text{with} \quad 0 \le \theta \le \pi.$$

The analysis is limited to α in the range $0 < \alpha < \pi$. Then $B > 0$, and A is positive or negative according to whether $\kappa/(1 - \tau) > \cos\alpha$ or not. Corresponding to whether A is positive or negative, θ will be in Quadrant I or II. With θ thus defined,

$$A\cos\omega t + B\sin\omega t = X\cos(\omega t - \theta)$$

where

$$X = \sqrt{A^2 + B^2} = \sqrt{\kappa^2 - 2\kappa(1 - \tau)\cos\alpha + (1 - \tau)^2}.$$

Therefore,

$$p = \frac{2\,m\,R\,T_C}{V_1\,[Y + X\cos(\omega t - \theta)]}.$$

It is easily seen then that

$$p_{max} = \frac{2\,m\,R\,T_C}{V_1\,[Y - X]} \quad \text{and} \quad p_{min} = \frac{2\,m\,R\,T_C}{V_1\,[Y + X]}.$$

Then the root mean cycle pressure $\overline{p} = \sqrt{p_{max}\,p_{min}}$ can be written as

$$\overline{p} = \frac{2\,m\,R\,T_C}{V_1\,\sqrt{Y^2 - X^2}}.$$

It can be checked that this is the exact same expression that is obtained by calculating the integral average pressure.

Using this, the instantaneous workspace pressure p may be finally written in terms of the average pressure $\overline{p}$ as

$$p = \frac{\overline{p}\,\sqrt{Y^2 - X^2}}{Y + X\cos(\omega t - \theta)}.$$

INDICATED WORK

Now the cyclic indicated work can be calculated as

$$W = \oint p\,dV = \int_0^{2\pi/\omega} p\,dV =$$

$$= -\frac{V_T\overline{p}\,\omega\kappa\sqrt{Y^2 - X^2}}{2\,(\kappa + 1)} \int_0^{2\pi/\omega} \frac{\sin\omega t}{Y + X\cos(\omega t - \theta)}\,dt.$$

The last integral can be worked out by well-known mathematical methods to yield this closed form expression:

$$W = \frac{\pi(1 - \tau)V_T\,\overline{p}\,\kappa\sin\alpha}{(\kappa + 1)\big[\sqrt{Y^2 - X^2} + Y\big]}.$$

FORCED WORK

The definition of the forced work of a cycle requires integrating the product of $p - p_b$ and dV over those portions of the cycle where they differ in sign, and only over those portions. The forced work integral $W_- = \oint[(p - p_b)dV]^-$ for a Schmidt gamma engine cannot be obtained as a closed-form expression. It is therefore necessary to employ numerical integration techniques. An identity for the negative part function which is convenient to use in the numerical integration is the following:

$$Z^- = \frac{|Z| - Z}{2}.$$

REFERENCES

Bejan, A. (1996a). *Entropy Generation Minimization.* CRC Press, Boca Raton.

Bejan, A. (1996b). "Entropy generation minimization: The new thermodynamics of finite-size devices and finite-time processes," *Applied Physics Reviews,* Vol. 79, pp. 1191–1218.

Buck, R. C. (1965). *Advanced Calculus.* McGraw-Hill, New York.

Curzon, F. L., and Ahlborn, B. (1975). "Efficiency of a Carnot engine at maximum power," *American Journal of Physics,* Vol. 43, pp. 22–24.

Finkelstein, T. (1960). "Optimization of phase angle and volume ratio for Stirling engines," Annual Meeting of the Society of Automotive Engineers, Paper No. 118C.

Gordon, J. M., & Huleihil, M. (1992). "General performance characteristics of real heat engines," *Journal of Applied Physics,* Vol. 72, pp. 829–837.

Hargreaves, C. M. (1991). *The Philips Stirling Engine.* Elsevier, Amsterdam.

Kirkley, D. W. (1962). "Determination of the optimum configuration for a Stirling engine," *Journal of Mechanical Engineering Science,* Vol. 4, pp. 203–212.

Kolin, I. (1972/1998). *The Evolution of the Heat Engine.* Longman, London (repr. by Moriya Press, 1998).

Kolin, I. (1983). *Isothermal Stirling Engine.* University of Zagreb Press.

Obert, E. F. (1960). *Concepts of Thermodynamics.* McGraw-Hill, New York, p. 200.

Rallis, C. J., and Urieli, I. (1976). "Optimum compression ratios of Stirling cycle machines," Research Report No. 68, University of Witwatersrand, Johannesburg, South Africa.

Ramelli, A. (1588/1976). *The Various and Ingenious Machines of Captain Agostino Ramelli of Ponte Tresa.* (Trans.) M. T. Gnudi with technical annotations and pictorial glossary by E. S. Ferguson, Baltimore, The Johns Hopkins University Press, The Scolar Press.

Reader, G. T. (1991). "Performance criteria for Stirling engines at maximum power output," *Proc. 5th International Stirling Engine Conference*, Dubrovnik, Paper No. ISEC 91069.

Ross, M. A. (1987). "A 90 cc inverted yoke drive Stirling engine," *Proc. 22nd Intersociety Energy Conversion Engineering Conference*, Philadelphia, Paper No. 879212.

Rudin, W. (1964). *Principles of Mathematical Analysis*. McGraw-Hill, New York.

Schmidt, G. (1871). "Theorie der Lehmannschen calorischen Maschine," *Zeitschrift des Vereines deutscher. Ingenieure*, Vol. 15, pp. 1–12, 97–112.

Senft, J. R. (1974). "Moriya—a 10-inch Stirling engine powered fan," *Live Steam Magazine*, December, pp. 10–12.

Senft, J. R. (1982). "A simple derivation of the generalized Beale number," *Proc. 17th Intersociety Energy Conversion Engineering Conference*, Institute of Electrical & Electronic Engineers, Paper No. 829273.

Senft, J. R. (1984). "A low temperature difference Ringbom Stirling demonstration engine," *Proc. 19th Intersociety Energy Conversion Engineering Conference*, American Nuclear Society, Paper No. 849126.

Senft, J. R. (1985). "Mechanical efficiency of Stirling engines—General mathematical considerations," *Proc. 20th Intersociety Energy Conversion Engineering Conference*, Society of Automotive Engineers, Paper No. 859436.

Senft, J. R. (1986). "A solar Ringbom Stirling engine," *Proc. 21st Intersociety Energy Conversion Engineering Conference*, Paper No. 869112.

Senft, J. R. (1987a). "Mechanical efficiency of kinematic heat engines," *Journal of the Franklin Institute*, Vol. 324, pp. 273–290.

Senft, J. R. (1987b). "Limits on the mechanical efficiency of heat engines," *Proc. 22nd Intersociety Energy Conversion Engineering Conference*, Philadelphia, Paper No. 879071.

Senft, J. R. (1988). "General results on mechanical efficiency of heat engines," *Proc. 23rd Intersociety Energy Conversion Engineering Conference*, American Society of Mechanical Engineers, Paper No. 889340.

Senft, J. R. (1989). "Charge pressure effects in kinematic Stirling engines," *Proc. 24th Intersociety Energy Conversion Engineering Conference*, Washington DC, Paper No. 899053.

Senft, J. R. (1991a). "Analysis of the brake performance potential of Crossley-Stirling engines," *Proc. 26th Intersociety Energy Conversion Engineering Conference*, Boston, Paper No. 910314.

Senft, J. R. (1991b). "Pressurization effects in kinematic heat engines," *Journal of the Franklin Institute*, Vol. 328, pp. 255–279.

Senft, J. R. (1991c). "An ultra low temperature differential Stirling engine," *Proc. 5th International Stirling Engine Conference*, Dubrovnik, Paper No. ISEC 91032.

Senft, J. R. (1992). "Mechanical efficiency considerations in the design of an ultra low temperature differential Stirling engine," *Proc. 27th Intersociety Energy Conversion Engineering Conference*, Society of Automotive Engineers, Paper No. 929024.

Senft, J. R. (1993a). *Ringbom Stirling Engines*. Oxford University Press, New York.

Senft, J. R. (1993b). "General analysis of the mechanical efficiency of reciprocating heat engines," *Journal of the Franklin Institute*, Vol. 330, pp. 967–984.

Senft, J. R. (1995). "A Stirling engine of the first degree," *Modeltec Magazine*, October, pp. 14–19.

Senft, J. R. (1996). *An Introduction to Low Temperature Differential Stirling Engines*. Moriya Press.

Senft, J. R. (1997). "Brake power maxima of engines with limited heat transfer," *Proc. 32nd Intersociety Energy Conversion Engineering Conference*, Honolulu, Paper No. 97413.

Senft, J. R. (2000). "Extended mechanical efficiency theorems for engines and heat pumps," *International Journal of Energy Research*, Vol. 24, pp. 679–693.

Senft, J. R. (2001). "The optimum Stirling engine geometry problem revisited," *Proc. 10th International Stirling Engine Conference*, University of Osnabrück, Germany, pp. 13–24.

Senft, J. R. (2002). "Optimum Stirling engine geometry," *International Journal of Energy Research*, Vol. 26, pp. 1087–1101.

Symon, K. R. (1971). *Mechanics*. Addison-Wesley, Reading.

Taylor, C. F. (1966). *The Internal Combustion Engine in Theory and Practice*. MIT Press, Cambridge, Massachusetts.

Walker, G. (1962). "An optimization of the principle design parameters of Stirling cycle machines," *Journal of Mechanical Engineering Science*, Vol. 4, pp. 226–240.

West, C. D. (1986). *Principles and Applications of Stirling Engines*. Van Nostrand Reinhold, New York.

INDEX